Climate Aotearoa

Climate Aotearoa

What's happening & what we can do about it

Edited by
HELEN CLARK

ALLEN&UNWIN
SYDNEY•MELBOURNE•AUCKLAND•LONDON

First published in 2021

Allen & Unwin
Level 2, 10 College Hill
Auckland 1011, New Zealand
Phone: (64 9) 377 3800

Email: info@allenandunwin.com
Web: www.allenandunwin.co.nz

83 Alexander Street
Crows Nest NSW 2065, Australia
Phone: (61 2) 8425 0100

A catalogue record for this book is available
from the National Library of New Zealand

ISBN 978 1 98854 763 3

Design by Megan van Staden
Cover image by Rachel Mataira
Set in 12/14.4 pt Baskerville
Printed and bound in Australia by Griffin Press

10 9 8 7 6 5 4 3 2 1

MIX
Paper from
responsible sources
FSC® C009448

The paper in this book is FSC® certified.
FSC® promotes environmentally responsible,
socially beneficial and economically viable
management of the world's forests.

CONTENTS

INTRODUCTION

Rt Hon. Helen Clark

This book of contributions on the climate crisis is being published as New Zealand, along with the rest of the world, continues to experience another crisis of global proportions — the Covid-19 pandemic.

The difference in nature of these two global crises is striking. Warnings of pandemic risk over the years have attracted far less publicity than those of the impact of climate change. When Covid-19 struck, however, its impact was rapid, dramatic and highly visible with far-reaching effects across human health, economies and societies. In the case of climate change, its impacts unfold over time as a slower onset disaster; if not addressed decisively, they will be far-reaching across the full spectrum of biological and non-biological systems.

Such has been the extent of social and economic carnage wrought by Covid-19 that governments have been forced to act ... In the case of the climate crisis, the same political dynamic and momentum have not been present, despite the staunch advocacy of many.

Furthermore, while effective measures for addressing the impact of Covid-19 have emerged quickly — such as physical distancing, hygiene, mask-wearing, testing and tracing systems, quarantine, treatment, and now vaccination, there is no parallel set of measures with immediate and dramatic impact for climate change mitigation — all necessary measures for that take time to implement and to gain traction.

Then, such has been the extent of social and economic carnage wrought by Covid-19 that governments have been forced to act — albeit with some doing so more effectively than others. In the case of the climate crisis, the same political dynamic and momentum have not been present, despite the staunch advocacy of many.

So, what of the book to hand? Does it provide a basis for optimism when it comes to tackling climate change in Aotearoa New Zealand? In the two opening chapters, Haylee Koroi and Jim Salinger set out current concerns within a longer-term perspective. Haylee points out that Māori have been dealing with the dynamics of the environment for centuries, and that New Zealand now has 'an opportunity to move into right relationship or whanaungatanga' aligned 'with the patterns and symmetry of our woven universe'. Jim Salinger reminds us that 'while differences in average temperature in our climate journey over centuries into the future may look small, they have a huge effect on the environment and the lives of plants and animals'.

Three further chapters — respectively by Rob Bell, Matt McGlone, and Simon Thrush and Andrew Jeffs — bring us up to date on the specifics of what science has to say about the impacts of sea-level rise, the effects of climate change on native flora, fauna, and biodiversity, and the trajectory for aquatic environments. What these chapters show is that human settlement in Aotearoa New Zealand has already inflicted huge damage on natural ecosystems, that climate change will add to this over time unless contained and mitigated, and that many of the actions we need to undertake for climate change mitigation are what we should do in any case to protect our environment.

After this essential work of setting the scene with the 'bigger picture', the book moves to more immediate issues, actions, and solutions. Rhys Jones and Kera Sherwood-O'Regan canvass the health impacts of climate change, warning that '(A)s climate change worsens, health inequities . . . are likely to be magnified'. Activists Jason Boberg and Sophie Handford add to this the perspectives of people with disabilities and young people respectively, underlining the interest and energy of wider communities in climate issues and their implications.

Then, on solutions, Jamie Morton, a science journalist at *The New Zealand Herald*, argues that individuals can make a difference, and that it is important that news media and scientists avoid doomsday narratives and focus on positive messages. Rod Oram takes a 'glass half-full' look

Among the reassuring features of this book is that it demonstrates that New Zealand has a community of scientists, journalists and activists that is passionate and well informed about climate change and willing to press the case.

at the issues posed by agriculture and food production, concluding that there is growing interest in regenerative agriculture among farmers, and that there are sound economic and environmental reasons for progressing down that path. Adelia Hallett of *Carbon News* notes that climate change is 'a classic tragedy of the commons' because no one owns the climate, and so no one person or country has the incentive to take care of it. She concludes that by the time this book is published, Aotearoa New Zealand will have just nine years to halve its emissions.

Among the reassuring features of this book is that it demonstrates that New Zealand has a community of scientists, journalists and activists that is passionate and well informed about climate change and willing to press the case. As well, action we take will be positive for our health and wellbeing and for our natural heritage. We want to preserve our indigenous forests. We want to prevent irreversible damage to coastlines, estuaries, wetlands, and other ecosystems. Our major cities are overrun with private vehicles and in need of containing urban sprawl. Our train network needs rescuing. Our primary industries are in almost all cases reaching and surpassing their environmental limits. As well, they are not yet adding sufficient value to our economy to support our lifestyle, and are in danger of becoming increasingly out of touch with changing consumer attitudes around the world. Our major tourist attractions were fully stretched pre-pandemic.

All this we know, but the need for change has to date been consigned to the slow lane of political possibility. Yet, as the Covid-19 crisis shows, when the 'team of five million' acknowledges the need to act, it does so to great effect. That is the sense of urgency with which the climate crisis now needs to be addressed.

New Zealand needs to revisit many features of its economic, social and environmental policy settings. The need to do so has often been foreshadowed during our pandemic experience. Both the pandemic and the climate crisis should be catalysts now for making the necessary changes.

I congratulate the authors on their contributions to this book and hope that New Zealand policymakers will draw on them as they design responses to the climate crisis. Like the pandemic, it has the potential to destroy our prospects, but it doesn't have to be that way. It is within our collective capacity to design and implement a more sustainable future.

In right relationship — whanaungatanga

HAYLEE KOROI
Indigenous sovereignty activist
and Māori public health advisor

Ka tū tonu Whakarongorua,
E whakarongo ana ki ngā tai e rua,
Ko te tai tama tāne e papaki mai nei ki Hokianga,
Ko te tai tama wahine e papaki mai nā ki Taumārere

Whakarongorua stands in perpetuity,
And I listen to the two tides
The tide of the male aspect which beats at Hokianga
And the tide of the female aspect which laps at Taumārere

(MEAD & GROVE, 2001)

It's March 2020 and I'm driving along the windy and pothole-filled roads of Te Hiku o Te Ika,[1] Aotearoa. The sun is rising in the east, behind my shoulder as I find my way across from Ōkaihau to the last winding strands of the Hokianga. Mataitaua marae nestles itself there, between the Utakura river, the shelter of the towering mountain Whakarongorua, and the dusty road heading to Horeke. These tireless landmarks are what I mostly remember about this place. I haven't been back here for any significant kaupapa in at least two years. We live in Auckland city, three hours away.

I finally arrive, pulling into the gravel carpark. I take a moment to compose myself. There's something about entering a marae that requires consideration. I try to see from afar what is happening inside. No luck, though; I'll just have to go in. I find a seat near the middle and look around to see if I know anyone. It's immediately evident to those in the room that I'm not haukainga. 'Ko wai koe?' they ask, a fairly typical question within the Māori world. In English this translates to 'Who are

---

1   Te Hiku o Te Ika a Māui denotes the tail of the fish (Te Ika a Māui) which was caught by Māui-pōtiki. Te Hiku o Te Ika is occupied by five iwi: Te Rarawa, Ngāi Takoto, Ngāti Kuri, Te Aupōuri and Ngāti Kahu.

you?' and could even sound rude if taken out of context. The way to answer this question is not necessarily to tell them your name. Instead, you tell them who your parents are, your aunties and uncles, and who your grandparents are. Within seconds several people are reciting back to me their own lines of whakapapa, pointing at faded photographs along the wall. 'Your great-grandmother is the sister to my great grandfather,' says a cousin. A nearby uncle who has overheard our conversation recites to relatives, as they pass by, that this is so-and-so's great-granddaughter. Each strand of whakapapa illuminated feels like the weaving of my spirit back into the tight hold of the collective.

The privilege I feel being able to return home and, with a little bit of mending, find my whakapapa still intact is not at all lost on me.

The privilege I feel being able to return home and, with a little bit of mending, find my whakapapa still intact is not at all lost on me. After a morning of introductions, we are served hāngi and marinated mussels for kai. Everyone pushes the seafood my way, knowing how deprived I am as an 'Aucklander'. I haven't done a single thing to contribute so far, and I'm even shooed away from helping clean up in the kitchen. There's a niggling emotion of shame I feel in the way that I'm treated like manuhiri, but I decide instead to see it as a pōwhiri, a welcoming back to the source. I turn my shame into humility and I sit and observe. I observe as local placenames are recited, and iwi boundaries are outlined on a map projected along the front wall. I try to memorise the names and faces of those in the room. I listen to the histories that come out as a prelude to the speakers' impassioned conclusions. One would think these are a lawyer's closing testimonies, but no, we're in a wharenui talking about our mokopuna and a future that we can build with them in mind.

Someday, I'll be able to repay the privilege by taking up my own responsibilities to weave in the strands of river and mountain, dirt road and mokopuna, for those who show up at the waharoa unsure and sceptical of their standing place. But for now, this is where I stand, a teina in the warm presence of my whānau, a teina in the sheltering break of my mountain.

In many ways, I am only new here. I am still learning what it means to observe the way that the mountain Whakarongorua teaches. I have not spent nearly enough hours atop its peak to tell the mountain what is coming. Climate change is not new to the mountain; it has seen and endured every rising and falling tide. Rather, the mountain is a new conversation to the space of climate change. A forgotten elder. A quiet tuakana. An uncited source.

We have an opportunity, now, to move into right relationship or whanaungatanga with our ancestors, our mokopuna and with one another

We are the meeting place of the past and the future, which are moving cyclically through us all the time. We have an opportunity, now, to move into right relationship or whanaungatanga with our ancestors, our mokopuna and with one another — to restore a way of being that sits in alignment with the patterns and symmetry of our woven universe. The remainder of this chapter will be just that, an exploration of my own experiences of coming into right relationship with past, future and present.

Ko Tīnana te waka

Ko Tūmoana te tangata

Ko Kokohuia te maunga

Ko Huria te awa

Ko Tangonge te wai

Ko Te Uri o Hina te hapū

Ko Te Rarawa te iwi

Before we were 'Te Rarawa', we were known by many different names, at home on many different shorelines among the sea of islands[2] of Te Moana-nui-a-Kiwa.[3] My ancestor Tūmoana carried those names, navigating the vast moana on the voyaging canoe Tīnana, to the northwestern shores of Te Ika a Māui, at Tauroa Point, a place more commonly known as Ahipara (Waitangi Tribunal, 1997). Our wharenui, Hōhou Te Rongo, stands only a five-minute drive inland from there, in the settlement of Pukepoto. That is where this story and tātai whakapapa begin.

I'm told by a close relative that the name Hōhou Te Rongo, which means 'to bind' or 'to bring peace', derives from a time in which Poroa, a rangatira of Te Rarawa, had negotiated a peaceful end to war among the iwi of Te Rarawa and Te Aupōuri. On his deathbed, Poroa shared the words, 'Kia ū ki te whakapono, me aroha tētahi ki tētahi' (Piripi, 2012, p. 34). Translated, this means 'Hold steadfast to truth and nurture each other'. In his time, Poroa had laid the foundations of collaboration for hapū of Te Hiku o Te Ika. Before Poroa's passing prior to 1832, he bore witness to some of the first encounters between settlers and hapū, and could likely foresee the need for peace making over the coming era (Piripi, 2012).

---

2   'Sea of islands' is a term I first discovered in the works of Epeli Hau'ofa, who sought to define the expansiveness of the Pacific by the totality of our broad relationships with land and sea.

3   Te Moana-nui-a-Kiwa denotes the vast ocean of Kiwa, Kiwa being a prominent ancestor of many ocean peoples.

Poroa's nephew Panakareao would succeed Poroa's legacy as a rangatira within Te Hiku o Te Ika. He had a number of trusted relationships with missionaries, and solidified these bonds through marriage and the gifting of land. Missionaries learnt to speak te reo o te kainga and hapū members learnt to read and write via the Bible. By 1835 hapū and settler relations were well practised, albeit with a few hiccups. By 1840, documents including He Whakaputanga o Ngā Rangatira[4] (1835) and Te Tiriti o Waitangi[5] (1840) were signed in order to affirm the tino rangatiratanga of hapū, and enable powers to the queen to correct the unruly and poor behaviour of settlers at Kororāreka. Rangatira would have seen mutual benefit in establishing relationships with settlers, in particular within the trade and export industries which were booming among many northern hapū. Arama Rata and Tahu Kukutai (Kukutai & Rata, 2017) explain Te Tiriti o Waitangi as 'this country's first immigration policy'; an opportunity for hapū to extend their network of relationships while maintaining their tino rangatiratanga. However, it was not long before rangatira could see that power was shifting, as settler numbers rapidly increased.

Within Te Hiku o Te Ika, a Land Claims Commission was established in order to rectify any pre-Tiriti land transactions. A total of seven pre-Tiriti land claims were originally heard in the Kaitaia area in February 1843. It is recorded that Panakareao and other rangatira acknowledged these land transactions, intending to maintain the promises they had made to early settlers. However, by this time Panakareao had become suspicious and dissatisfied with the governor and stated that 'their hapū would resume any lands not granted to the claimants. In addition, they would not transact any more lands or allow any future interference by the Government' (Te Rarawa & The Crown, 2012, p. 35).

---

4   He Whakaputanga o Ngā Rangatira, also known as the Declaration of Independence, was the first treaty signed between hapū of the Confederated Tribes and the British Crown.
5   Te Tiriti o Waitangi refers specifically and only to the Māori text, which is not the same as the English text. It is this document which was predominantly signed by rangatira.

The processes which came out of the Land Claims Commission proved to be highly problematic, with land areas arbitrarily located, larger than originally agreed or ignoring clauses of joint occupancy, which were very common. Joint occupancy and other similar verbal agreements reinforced the idea that 'in Māori terms an agreement symbolised a human relationship which would change over time. They [Māori] did not understand the Pakeha concept of a binding in perpetuity . . .' (Rigby & John, 1989, p. 47). Ideas of permanence had no space within the ever-changing nature of whakapapa, which had to be consistently negotiated with, yielded to, shaped and directed in order to ensure the maintenance of good relationship with our human and more-than-human relatives.[6]

In 1857, during processes of land surveying, surplus lands at Pukepoto were subsumed under Crown title. This included a 685-acre land block, called the Tangonge block, that had been explicitly reserved for local hapū by missionary Reverend Joseph Matthews. My great-grandfather, Tīmoti Pūhipi, who would later pick up the mantle of leadership within Pukepoto, petitioned multiple commissions over successive years to have the land returned. Those presiding over the case decided either that the land was not Matthews' to reserve, or that there was insufficient evidence to show that a transaction of that property had taken place. Over the 1940s to the 1960s, while the Crown had allocated occupation licences for the Tangonge block to settlers, Pukepoto hapū members remained. The last of those who were considered 'squatters' were removed in the 1960s by the Native Department, claiming poor living conditions (Te Rarawa & The Crown, 2012).

What began as an invitation to extend hapū relationships to settlers

_____

6   More-than-human was a term brought to my attention by a friend, Hana Burgess, and sourced
     from Kim Tall Bear, a descendent of the Cheyenne and Arapaho tribes of Oklahoma. In the
     context of this chapter, the term 'more-than-human' denotes those who precede the arrival of
     humans in the whakapapa of creation from the separation of Ranginui and Papatūānuku down
     to the creation of atua and their offspring — fish, birds, lizards, etc.

(now tangata tiriti[7]) through the gifting of land left our hapū landless. Without access to ancestral lands, the social organisations of whānau, hapū and iwi were severely disrupted. The British Crown had usurped power, subjugating the tikanga of Pukepoto hapū to the shadows, and stripping hapū of their own authority and agency. It was clear that a new order was taking root within the land, undermining concepts of whakapapa which informed Māori conceptions of relationship with the human and more-than-human worlds.

This new faux-order called settler-colonialism gave power to ideologies of white supremacy, capitalism and patriarchy (hooks, 1997). More often than not, upper-middle-class white men were central in the establishment and maintenance of these power structures (Pihama, 2018), which were shaped by colonial laws, policy and cultural norms. This is clear in the very first Act passed through Parliament, the New Zealand Constitution Act (1852), which explicitly qualified voters as 'every man of the age of twenty-one years or upwards, having a freehold estate in possession'. As hapū held lands collectively and not under freehold estate, this prevented Māori men from voting by default. Māori women sat at the opposite end of the spectrum of colonially defined power, and were treated as inferior and subhuman.

These imported colonial ideologies justified the domination and exploitation of whānau, hapū and iwi, and by extension our more-than-human relations. This was indicative of broader global patterns of Indigenous erasure and land and water exploitation by the British Crown. In more recent times, these colonial ideologies have become so deeply embedded in our collective psyche that they have become almost imperceptible. However, our local, national and global histories can return us to the truth of who we are, where we have come from and, more importantly, where we are going. As bell hooks says, 'At the heart

---

7   Tangata tiriti denotes those who are bound to Aotearoa through Te Tiriti. In contrast, tangata whenua
        denotes those bound to Aotearoa through enduring relationships to local lands and waters.

of justice, is truth telling' (hooks, 2000, p. 33). Imagining futures beyond colonial domination is not only important but imperative, because it is there, and only there, that a just climate future is possible.

I offer some questions below as a starting point that may help guide you to perceive colonial ideologies and their distinctness more clearly. These questions are important in offering up space for us to perceive alternative logics, which may be more helpful in directing and aligning our viewpoints with the work we are doing.

- Before you were the nationality you now identify with, by what name were your ancestors known?
- Who were the first of your ancestors to land on the shores of Aotearoa?
- Where within Aotearoa did your ancestors first land?
- What were the truths or beliefs that your ancestors held?
- How did your ancestors' beliefs compare with the beliefs of those already here?

Ano te mara taro a Taiawarua, me nga koroi o Hotu, ka puhia te hau ka Tangonge noa.

Behold the taro gardens of Taiawarua and the kahikatea groves of Hotu, when the wind blows the taro leaves undulate and ripple.

(HONGI, 1930; GRAHAM, 1991; HENWOOD, 2016)

Tangonge was the name of the lake located in Pukepoto; as outlined above, its name refers to the sound of the wind as it blows through the leaves of the taro and the berries of the kahikatea. As well as providing fertile soils and water for nearby gardens, Lake Tangonge was used regularly as a place to gather tuna, pipi, freshwater mullet called kanae raukura, freshwater mussels called kāeo, and a particular type of short-sighted duck (Ministry of Justice, 2016, p. 38). On all sides of the lake, various hapū would exercise their rights to access the lake and its fruits daily. The centrality of the lake within daily life was clear, and is recorded in particular by one local, Herepete Rapihana, who said, 'We relied upon the lake in former times for our food supplies. Our people had their homes along the edge of the lake whilst engaged in fishing or hunting' (Ministry of Justice, 2016, p. 38). He also vividly recalls significant changes in the lake due to heavy rains, or high winds which would fell trees across the river's mouth. He talks about a particular instance 'when willows partly blocked the Kaitaia River, and more water was diverted down the Mangawhero Stream into the lake. The Maori, alarmed at the rise of the lake, then dug a drain to re-open the Waihou Channel on the north side of the lake' (Ministry of Justice, 2016, p. 38). Locals were committed to reducing the swollen lake levels and retaining the lake as a food pantry.

However, in 1913 this incident was taken advantage of, and drainage operations were carried out by the government under the Kaitaia Land Drainage Act (1913). The lake and its entire fisheries were destroyed. The lake was drained again in the 1920s under the Swamp Drainage Act (1915) to open up new land for returning soldiers and dairy farming. Many of the local hapū were against plans to drain the lake; not only would they not benefit from its draining in any way, but they would also be significantly worse off, having lost a vital food source. The most they could do was get paid to do the labour of digging the deep trenches.

The Swamp Drainage Act and other Acts which were employed to similar effect had huge implications across a number of lake-wetland sites including Kaitaia, Hauraki (Waitangi Tribunal, 2006) and the Rangitaiki river (Waitangi Tribunal, 1999). Within each of these localities lakes and wetlands were drained to extend useable lands for settlement, in particular for returned (non-Māori) soldiers and intensive farming endeavours. Hapū saw these bodies of water as an important food source and vital element within the local ecosystems. These drainage schemes contributed to the rapid loss of Māori lands, while hapū were never consulted. The drainage of these bodies of water has contributed to ongoing issues of water filtration, silting, low water levels and native species decline.

The destruction of local ecosystems which occurred nationwide after the British Crown usurped power from hapū was widespread. The loss of access to ancestral food systems and the destruction of ancestral sites has had devastating implications for the wellbeing of hapū. Well-integrated lakes and swamp ecosystems containing harakeke, raupō and tī kouka, which would collectively act as a significant carbon sink, were destroyed across Aotearoa. More often than not, these carbon sinks were replaced with significant carbon sources, such as intensive farming.

If we had listened to the first cries of injustice at the hands of settler-colonialism, we would not have had to wait until settler populations were in harm's way before deciding to act.

When we look at this very specific practice of lake–swamp drainage we are looking at only a glimpse of the full menu of extractive processes undertaken by the Crown, including forestry, oil drilling, mineral mining and commercial fishing. The cumulative impacts of all of these sites of social and environmental exploitation and injustice invariably paint a clear picture of where climate change has originated — within the same white-supremacist, capitalist, patriarchal systems that were mentioned earlier. The trouble is, these same colonial ideologies are being reproduced within spaces which purport to be leading solutions on climate change. This is conveyed in how climate change is framed, who the dominant voices on climate change are, and just *when* climate change became a mainstream point of discussion.

If we had all cared enough about the wellbeing of hapū, we would have caught the early localised symptoms of climate change long before they became manifest within the global environment. If we had listened to the first cries of injustice at the hands of settler-colonialism, we would not have had to wait until settler populations were in harm's way before deciding to act. If we had realised sooner that climate change is about so much more than the science of carbon, and is about justice, love and community, we would not be *still* struggling to take the level of action necessary to avoid such an *avoidable* disaster.

Real change will require nothing less than huge personal and collective paradigm shifts in how we view and relate to the world around us. I offer the concept of whakapapa as a guiding logic, which has lived here since time immemorial. It speaks to the ultimate reality of our interrelatedness and interdependence as both human and more-than-human beings. It will require us to think beyond our human desires to the wider web of relationships in which we exist. You can start where you are right now, illuminating and building on your closest relationships:

- What are or were some historically significant landmarks within your area?
- What were some of the previous uses of the land you now reside on?
- Who are the local hapū in your area?
- What are the Māori names of the places you reside in?

Whakapapa weaves all of existence together into an ever-expanding web of intimate relationships, forming the basis of Māori ways of being, knowing, and doing. Connected by our origin stories, whakapapa reveals that this web of relationships is whānau, existing in a state of whanaungatanga. With whakapapa, past and future generations, our mokopuna and tūpuna, are intimately related. Mokopuna are reflections of tūpuna. Whakapapa can explain the origins, positioning, and futures of all things. Any conversation about the future, is inherently a conversation about whakapapa. With this, any conversation about the future is inherently about our relationships with our mokopuna. Here, the future is not something unknown and separate from us, but something that we are intimately related to all the time. — Hana Burgess and Te Kahuratai Painting, Whose Futures? *(2020)*

He tamaiti akona ki te kāinga,
tū ki te marae, tau ana.

A child who is taught well at home
will stand tall on the marae.

In 1907, the same year that New Zealand changed from a colony to an independent dominion, the Tohunga Suppression Act (1907) was passed. The preamble of the Act reads, '[Tohunga] practise on the superstition and credulity of the Maori people by pretending to possess supernatural powers in the treatment and cure of disease, the foretelling of future events, and otherwise, . . . to the injury of themselves and to the evil example of the Maori people generally.' This assessment of the work of tohunga is again rooted in imported colonial ideologies. The Tohunga Suppression Act, alongside the Native Schools Act (1858) which banned Māori language in schools, contributed to the devastation of mātauranga-a-hapū, and likely led to the subsequent disavowal of Māori from their spiritual selves. The Tohunga Suppression Act was not repealed until 1962, the year my father was born.

Despite this, numerous people, including many haukainga, were able to maintain local histories and knowledge for their whānau and hapū. I've been fortunate in recent years to meet a number of those people, and in November 2019 I was able to return to Te Hiku o Te Ika to speak with them on the kaupapa of climate change. I was preparing for a trip to the United Nations Forum on Climate Change (COP25), which would be held in Spain that December. It was daunting to think about the sheer size and power of a forum such as the United Nations. Similar to the whakataukī mentioned above, I knew that to stand confidently at the United Nations I would have to be well informed by those from home and the knowledge that they had attained through years of experience. I was fortunate to talk to a number of both close and distant relatives who were engaged in community work and who, for all intents and purposes, should be considered climate leaders within their own right.

I met and spoke with Waikarere Gregory (Te Rarawa, Te Aupōuri who is living in Kaitaia. She is a leader in many ways within her community, supporting the restoration of Tangonge and holding a lot of historical knowledge of Pukepoto for future generations. I've had the privilege of listening to and learning from her a few times while I've been back home. On this particular occasion, however, we spoke about the EcoCentre Kaitaia, where she works occasionally, and the local time-banking system, of which she is a member. Time-banking is an alternative to a money economy, and is based on the idea that all people have something of value to contribute to the community. People accumulate time credits (one credit is equivalent to one hour of work), and these acquired time credits are exchanged for the resources, skills or time of other people. This means that those who are doing work which is often unrecognised, for example kaikaranga, volunteers or elderly people who can give in very particular ways, can be recognised for their work and the value that it brings. The Kaitaia Time Bank currently has around 200 registered participants, and there are a number of other time-banking systems throughout Aotearoa, based on the book *No More Throw-away People* by Edgar S. Cahn (2000). When I asked how the time-banking system had changed the community, Waikarere said that the system was 'largely preaching to the converted'. Community care was an inherent and necessary part of community life, but she also saw an increase in the participation of the elderly, who are often isolated from their communities. It is hoped that the digital administration of the system will eventually become obsolete, once whanaungatanga becomes an inherent part of the community, as it once was.

I also met with a mentor, Wendy Henwood (Te Rarawa), who worked as a community researcher at the intersections of public health and environmental leadership. Wendy and a group of researchers did a two-year research project on 'Enhancing Drinking Water Quality in Remote Communities' (Henwood et al., 2019). This was an especially pertinent topic in Te Hiku o Te Ika, considering regular patterns of drought during hot summers. The research team wanted to understand what the impacts of climate change were on household drinking water. 'If houses haven't got drinking water, we don't actually have communities, and we don't have people,' said Wendy. Aside from quantitative measures, Wendy shared that, 'Our people weren't fazed by climate change. They all had stories of, "hey, it's been changing all my life". The change didn't worry them because "we're just adaptable people". The possibility of losing access to household drinking water was scary for some people, but then they were practical and said "back in the day we relied on puna, we didn't have tanks and spouting. We had different puna for different purposes, one for the garden, cowshed, household."' Wendy spoke about whānau at Pawarenga who were very worried about sea-level rise, as their marae, which sits on a low-lying coastline, is already experiencing problems because of this. Wendy said that the costs involved in moving a marae worried a lot of people. This story shows, contrary to popular narratives, that local communities are aware of climate change and have been adapting to local climate change for many years.

More broadly, what these conversations with Wendy and Waikarere made evident was the importance of locating climate solutions within the context of the local community. Our communities are already leading local responses to climate change — they just happen to be doing so without the recognition or the resources that they need. Rural communities in particular don't have the luxury of infrastructure to delay the implications of climate change, so they feel the effects almost immediately. It therefore made a lot of sense that during my travels I found that communities were already practising food sovereignty, growing, foraging, hunting or diving for their own foods. They also practised water preservation, and community interdependence was a well-integrated reality for many rural communities.

When we look at those who get to define climate change, and therefore solutions to climate change, often they are people who are completely removed from the local communities discussed.

When we look at those who get to define climate change, and therefore solutions to climate change, often they are people who are completely removed from the local communities discussed. In many instances, national climate policies assert generic climate solutions across highly diverse communities, without having to take responsibility for the implications of these policies on the everyday lives and life ways of local people. A few friends of mine have talked about the implications of pine monocrops which have dried up nearby rivers, destroying toheroa beds, degrading soil for years or leaching acid into waterways, killing local eel populations. Yet the government has promoted and made significant financial investments into pine forestry as a part of its Emissions Trading Scheme (ETS). This tactic responds to our most shallow conceptions of climate change, as *merely* an issue of carbon emissions. Solutions like these may spare us in the very short term, but the harm of underlying exploitative behaviours will only re-establish themselves in other pressing environmental issues, including rapid loss of biodiversity, water degradation, plastic overconsumption, etc. Local communities must be afforded the decision-making power to do the work that they know is necessary to maintain the health and wellbeing of the local ecosystems they occupy. They must be better supported and resourced to ensure that local peoples are well fed and sheltered, and that they have the capacity to continue to carry out the work they are already doing.

In a similar way that we understand the whakataukī 'He tamaiti akonga ki te kainga' as a call to ground ourselves in the tikanga of home, so too should we use this as a guiding principle when thinking about effective climate action, by ensuring that the work of local communities grounds our overall regional and national policies.

- Do you know your immediate human neighbours?
- Do you have trusted relationships within your immediate local community?
- Are the people in your neighbourhood sufficiently fed and supported?
- How is the wellbeing of your more-than-human neighbours?

Anō te pai, te āhua reka o te noho tahi o te teina
me te tuakana i runga i te whakaaro kotahi.

Behold, how good and wonderful to sit together,
teina and tuakana, in a common thought.

I also met with a friend, Piripi Tautari (Te Arawa, Tainui, Ngāpuhi, Te Roroa, Te Rarawa), and his whānau. They maintain a garden and an orchard, and also keep a few chickens on his family land in Waiōmio. Beyond his garden he hunts and fishes to ensure his family is fed.

Piripi lives in the world of atua. Where tomatoes self-seed, clogging the garden pathways, he gives thanks to Haumie-tiketike, the atua of uncultivated foods. Haumie-tiketike is Piripi's favourite atua, because he gives of his gifts for free, and asks for little in return. He also believes that the fruit is always sweeter, because it germinates when the conditions are most ideal for the seed, not when a human decides it's a good time. The kūmara and Māori potatoes which have sent tender leaves through the mulch sit within the realm of Rongomātane, the atua of cultivated foods. Piripi has a number of ahurei, large mauri stones which represent the realms of these various atua and maintain the mauri of their associated realms.

Piripi is also a healer. While I'm there he hears me coughing, and prepares a tea of kūmarahou and tītoki which will help to break down mucus. Of course, the ahurei within this section of Piripi's garden belongs to Tāne, the atua of the forest.

Piripi reckons he could talk forever about these things. I believe him. Piripi lives in a world far removed from the worlds that many of us have grown up in. His language and words depict a relationship with atua that is completely at odds with the narratives of dominion over our more-than-human relatives that we are raised in.

One thing in particular sticks with me as I pack my things to leave. Piripi talks about the birds, who are tuakana, elder siblings to humans. He says, 'You have to get up and do your karakia before the birds, because the tuakana always gets the last say.' Piripi is referring to a tikanga often found within the context of whaikōrero, in which the tuakana (eldest sibling) always closes. Those things which seem obvious to Piripi completely shake the foundations of what I know and have known. In my puku,[8] his words fall into place. It's a long, beautiful and often sorrowful journey to remember our more-than-human relatives,

---

8   The puku or stomach is often referred to as the knowledge and emotional centre of the body. There are many words that connect Māori emotions to the region of the stomach. This includes words such as pukumōhio — to know via the stomach; pukukata — to laugh from the stomach; pukuriri — anger centred in the stomach.

but each time I feel more and more at home.

Rereata Makiha (Te Mahurehure, Te Arawa) is another great mentor of mine who I have had the greatest privilege of learning from. Rereata is one of the few remaining students of the Ngāpuhi Whare Wānanga. His understandings of whakapapa and maramataka and the implications of lunar phases on human and more-than-human behaviour would make him a tohunga by anyone's standards. Rereata talks about 'ngā tohu o te whenua' and 'ngā tohu o te rangi', or indicators upon the land or sky. These tohu tell us that a new phase has begun, and allow us to respond accordingly. Rereata talks about matiti muramura, the third summer phase indicated by the blossoming of pōhutukawa and rātā. Rehua is known as the prominent star of this period, or 'te tohu o te rangi'. During this time kina are ready to be harvested.

Rereata isn't a scientist in the formalised western sense, but he engages in the method of observation as his tīpuna did for thousands of years. He often relates the story of erosion and deposition to the pakanga at Te Paerangi,[9] where Tāne and Tangaroa fought. In particular, Rereata talks about the whakapapa of Moana tū i te Repo, the swamp maiden, whose role was to mediate peace between Tāne and Tangaroa. The swamp maiden filters sediment and toxins from the water as it passes from land to swamp, and out into our waterways. Rereata talks about how the progressive loss of swamps in particular across Aotearoa contributes hugely to many of our water-degradation issues. Over the past two to three years we have seen a slow resurgence of whakapapa knowledge and maramataka in common Māori vocabulary thanks to the hard work of people like Rereata.

Piripi and Rereata both understand and exemplify the concept outlined in the whakataukī, 'Anō te pai, te āhua reka o te noho tahi o te teina me te tuakana i runga i te whakaaro kotahi.' This whakataukī, which

---

9   Te Paerangi as explained to me by Rereata is short for 'Te Pae o Te Rangi' — a place, a kind of platform where two atua come into contact with one another.

is written on the walls of Te Rarawa marae, calls us into partnership, unity, collaboration. It alludes to a 'tuakana–teina' relationship, equivalent to that of 'older sibling, younger sibling'. Tuakana, by definition, have spent more time within the world, and understand more clearly the limitations of our human and more-than-human relationships. Tuakana lead and guide, because they have an inherent responsibility to use their knowledge to ensure the continuation of whakapapa. Hapū are tuakana within the context of their ancestral lands and waterways.

Relatively speaking, tangata tiriti are teina within the whakapapa of Aotearoa, the very youngest siblings.

Hapū who have maintained their relationships with their more-than-human relatives, despite years of colonial violence, continue to pass down ancestral knowledge to their mokopuna.

Relatively speaking, tangata tiriti are teina within the whakapapa of Aotearoa, the very youngest siblings. This means tangata tiriti may not have the experience or expertise necessary to respond to the broad and inter-related needs of all of our human and more-than-human relatives. The contemporary social and environmental conditions of inequity within Aotearoa, in my view, prove that assumption to be true.

When we look at the broader global context, in the short time that European nations have colonised a majority of the globe, they have led humanity to the precipice of an extinction-level event. Conversely, we know that Indigenous people continue to protect 80 per cent of the world's remaining biodiversity within Indigenous lands.

As stated above by the whakataukī, it is imperative that tuakana and teina continue to work together to imagine a future that we can build with our mokopuna in mind. Tuakana lead, guiding teina within the boundaries of safe co-existence with all of our relatives. In order for Indigenous communities to do that work, it is imperative first and foremost that colonial systems devolve power and resources back to hapū. We cannot reasonably expect Indigenous communities to lead such huge undertakings without the resources necessary to feed their people, put a roof over their heads and ensure they have sufficient healthcare. Once this power dynamic has been corrected, and hapū tino rangatiratanga restored as per Te Tiriti o Waitangi, we may have a real chance — a chance not only to collectively get through to the other side of this climate crisis, but to do so while (as my waka-ama coach often says) 'still liking one another'; still committed to being in right relationship (whanaungatanga), as descendants from many different lineages, brought to this place and time to see this challenge through together.

As tuakana, hapū will not *always* have all of the answers, and that is OK. In the whakapapa of creation, we as hapū are also teina. Teina to the birds, the trees, the rocks, even the waters, who have known this place much longer than we have. So we noho with our tuakana, as the whakataukī instructs. We sit with humility and observe the way that atua exist within the world, and we temper our actions accordingly. To live in interdependence with atua is not optional; as we have seen within our pūrākau, when Māui tried to gain immortality through Hine-nui-te-po, he was crushed and died. If, as teina, we continue to disregard the boundaries set by atua, it will be to our own peril.

We need not repeat the lessons of the past; they are already known. There are new lessons to be had, always, within the ever-changing nature of whakapapa. Whakapapa, which must be consistently negotiated with, yielded to, shaped and directed in order to ensure we are in right relationship with our human and more-than-human relatives. Climate justice is not a place of arrival; rather, it is a way of seeing and being *of* the world.

- Who are the human tuakana within your field of work?
- Who are the more-than-human tuakana within your field of work?
- Have you engaged in the practice of noho teina?

When the rains are heavy the waters rise, and you can sometimes see the edges of Tangonge reach outwards to the places and edges it has known in times past. The waters swell closer to where my own grandmother once wandered, and where ancestors before her did the same. I am the meeting place of those who have walked the lake's edge before me, and those who will walk after I am gone.

However, the whakapapa of Tangonge isn't just about me or my ancestors any more. It is about many of us. Together we have an opportunity to move into right relationship or whanaungatanga with our ancestors, our mokopuna and with one another — to restore a way of being that sits in alignment with the patterns and symmetry of our woven universe. The remainder of this story or tātai whakapapa must be just that: an exploration of our experiences of coming into right relationship with past, future and present.

QUESTION SUMMARY

The questions below, which have also been posed throughout this chapter, aim to guide us through a process of coming into right relationship with our whakapapa. We might perceive ourselves as individuals, separate from our ancestors and the legacy they have left for us, but this is quite simply an illusion. Being in right relationship requires that we understand, as much as possible, how it is that we have arrived at our current standing place; also, how the systems around us either reinforce or deny our right to stand, and how in turn this right is extended to others. Only then can we truly take responsibility for the roles that we must play in bringing about climate justice, and ensuring the continuation of whakapapa into the future.

- Before you were the nationality you now identify with, by what name were your ancestors known?
- Who were the first of your ancestors to land on the shores of Aotearoa?
- Where within Aotearoa did your ancestors first land?
- What were the truths or beliefs that your ancestors held?
- How did your ancestors' beliefs compare with the beliefs of those already here?
- What are or were some historically significant landmarks within your area?
- What were some of the previous uses of the land you now reside on?
- Who are the local hapū in your area?
- What are the Māori names of the places you reside in?
- Do you know your immediate human neighbours?
- Do you have trusted relationships within your immediate local community?
- Are the people in your neighbourhood sufficiently fed and supported?
- How is the wellbeing of your more-than-human neighbours?
- Who are the human tuakana within your field of work?
- Who are the more-than-human tuakana within your field of work?
- Have you engaged in the practice of noho teina?

Glossary

AHUREI large mauri stones which represent the realms of various atua and maintain the mauri of their associated realms

ATUA guardians of natural realms

HAPŪ subtribal grouping

HAUKAINGA local people of the marae

IWI tribal grouping

KAI food

KAIKARANGA someone who calls visitors into a protected space, invoking ancestors

KARAKIA incantation, chant or recital of whakapapa or acknowledgement of atua, tipuna, etc.

KAUPAPA topic of discussion

MARAE central gathering place of local hapū, also physically outlined via the atea or courtyard and relevant buildings

MARAMATAKA lunar calendear

MATAURANGA knowledge of or way of seeing the world, world view

MĀTAURANGA-A-HAPŪ knowledge located in the context of hapū

MAURI life force or life essence

MANUHIRI visitor or guest entering into a space that is protected by a particular group, including whānau, hapū and iwi

MOANA ocean

MOKOPUNA descendants

NOHO to sit, to be still, to contemplate

PAKANGA war

PŌWHIRI welcoming protocol for visitors

PUKU stomach

PUNA spring

PŪRĀKAU creation stories

RANGATIRA leader

TĀTAI WHAKAPAPA a particular lineage of whakapapa or genealogy

TINO RANGATIRATANGA absolute authority and agency as inherited through decendence from atua

TOHUNGA skilled person, chosen expert, priest, healer — a person chosen by the agent of an atua and the tribe as a leader in a particular field because of signs indicating talent for a particular vocation (www.māori dictionary.co.nz)

TE REO O TE KAINGA the language of the whānau and hapū unit, which dialectically varies across local areas and regions

TEINA the youngest, the least experienced or the entity that is still growing within kaupapa or a particular body of knowledge

TUAKANA the eldest, most experienced, or most advanced entity within a kaupapa or body of knowledge

WAHAROA entrance or doorway onto the marae

WHAIKŌRERO a formal speech

WHAKAPAPA whakapapa denotes a web of relationships originating from Papatūānuku and Ranginui. From these two descend all life, from atua to tangata, through genealogical lines of descent.

WHAKATAUKĪ proverb or saying

WHĀNAU smallest unit of Māori social organisation

WHARE WĀNANGA traditional houses or spaces of learning esoteric knowledge

WHARENUI ancestral meeting house

Bibliography

H. Burgess & T. K. Painting. 'Māori futurisms'. In Anna Maria Murtola (Ed.), *Whose Futures?*. Tāmaki Makaurau/Auckland: Economic and Social Research Aotearoa, 2020.

M. Graham. *Whakapapa and Historical Notes and Accounts*. Wellington: Alexander Turnbull Library, 1991.

W. M. Henwood. Ko Tāngonge Te Wai: Indigenous and technical data come together in restoration efforts. *EcoHealth*, 2016, vol.13, pp. 623–632.

W. Henwood, T. Brockbank, H.M. Barnes, E. Moriarty, C. Zammit & T. McVreanor. Enhancing drinking water quality in remote Māori communities. *MAI Journal*, 2019, vol. 8, no. 2.

H. Hongi. *Ancient Maori history: recollections of a rambler*. Wairoa: Wairoa Star Print, 1930.

b. hooks. Cultural Criticism & Transformation (M. Patierno, S. Jhally & H. Hirshorn, Eds). Media Education Foundation, 1997.

b. hooks. *All About Love*. New York, NY: William Morrow, 2000.

T. Kukutai & A. Rata. 'From mainstream to Manaaki: Indigenising our approach to Immigration'. In D. Hall (Ed.), *Fair Borders? Migration policy in the twenty-first century* (pp. 26–44). Wellington, New Zealand: Bridget Williams Books, 2017.

H. M. Mead & N. Grove. *Ngā Pēpeha a Ngā Tīpuna*. Wellington: Victoria University Press, 2001.

Ministry of Justice. *Judge Acheson, The Native Land Court, and the Crown*. Wellington: Ministry of Justice, 2016.

L. Pihama, 'Colonization and the importation of ideologies of race, gender, and class in Aotearoa'. In E. A. McKinley & L. T. Smith (Eds), *Handbook of Indigenous Education*. Singapore: Springer, 2018.

H. Piripi. *Wai 45: Brief of evidence of Haami Piripi on behalf of Te Runanga o Te Rarawa*. Wellington: Pacific Law Limited Barristers & Solicitors, 2012.

B. Rigby & K. John. *Muriwhenua Land Claim (Wai-45)*. Wellington, New Zealand: Waitangi Tribunal Division,1989.

Te Rarawa & The Crown. *Te Rarawa Deed of Settlement*, 2012.

The Waitangi Tribunal. *Muriwhenua Land Report (Wai 45)*. Wellington: GP Publications, 1997.

The Waitangi Tribunal. *The Ngati Awa Raupatu Report*. Wellington: Waitangi Tribunal, 1999.

The Waitangi Tribunal. *The Hauraki Report Volume III, WAI 686*. Wellington: Waitangi Tribunal, 2006.

PART 1

The Science

Climate: Recent past, present and future

DR JIM SALINGER
Scientist and climate change researcher,
Victoria University of Wellington

O ur climate has changed already, and will continue to change during the twenty-first century. Since the nineteenth century there has been an upsurge of anthropogenic global warming — that is, the long-term increase in the average temperature of the Earth's surface because of human industry and agriculture. Along the way there have been bumps in the warming road, because of climate variability — events such as volcanic eruptions causing temporary dips, and the ups and downs of the La Niña and El Niño weather patterns — but in general the trend is definitely upwards.

Given that this trend seems likely to continue for some time, scientists have been looking at scenarios of what our climate could do for the remainder of this century, and the dramatic impacts of this on Aotearoa and its inhabitants. We are likely to see changes in the frequency of extreme events such as heatwaves and drought, and huge effects on flora and fauna, freshwater, New Zealand's surrounding marine resources, livestock, and arable and fruit crops.

Nineteenth-century climate and end of the 'Little Ice Age'

In the nineteenth century, the globe was much colder than it is now. The world was experiencing the end of the Little Ice Age, a time of cooler climate over much of the planet which lasted about 250 years, from around 1600 to around 1850. The first reliable thermometer measurements in New Zealand, taken in the late 1860s, show temperatures that were close to 1°C below average temperatures today. We were particularly fortunate that the first Pākehā government scientist, Sir James Hector, established a world-class climate network here in the late 1860s, with recordings made throughout the colony using precision instruments, producing high-quality measurements.

These nineteenth-century climate records show a climate 0.8°C cooler than the 1981–2010 period, capturing the end of the Little Ice Age. This was an environment where our cooler climate produced much more severe winters in the South Island, and evidence of this cooler climate is contained in Robert Gilkison's *Early Days in Central Otago* (first published in 1930):

> *The winter of 1862 proved exceeding hard and cold and was especially felt by those who came down from warmer climates and unused to tented life. So keen was the frost that Lake Waihola [near Dunedin] was frozen, and skaters were able to traverse some 3 miles of ice, a state of things never known since.*

Evidence of the earlier impacts of the cooler landscape and greater blanket of snow and ice in the mountainous areas of the world comes from the explorers who mapped the glaciers in Europe, Asia, the Americas and the Southern Alps of New Zealand. The most recent maximum extent of our glacier mantle, inducing the clouds which cloak the Southern Alps and give this country the name 'Land of the Long White Cloud', occurred in the 1890s. At this time the iconic Fox and Franz Josef glaciers were about 3–5 km longer than they are today.

The cooler climate is further illustrated by the location of the old Ball Hut, built in 1891 on the Tasman Glacier near Aoraki/Mount Cook. The hut was built at the same level as the moraine terrace (formed by rocks, sand and clay carried by the glacier). The glacier has subsequently melted (a process called downwasting), particularly rapidly in the last part of the twentieth century. By 2019, the surface of the glacier was 210 m below the moraine terrace formed in the 1890s.

Warming in the past seven decades

Global mean temperatures showed little change up to 1920, when the climate began to warm up. There was a pause in the 1950s and 1960s, when increasing air pollution from industries blocking out sunshine in North America, Europe and eastern Asia may have affected the heating up of the northern hemisphere, before the increases in temperature resumed. In New Zealand from 1900 until 1940, mean annual temperatures remained about 0.8°C cooler than the 1981–2010 climatology period. By the mid-1950s temperatures had increased by 0.5°C, then these trended upwards by 0.9°C by 2019, largely due to anthropogenic global warming (see Figure 1).

Changes in extreme temperatures have also occurred, associated with the warming climate of New Zealand. The average number of summer days (in December, January and February) with recorded temperatures of 25°C or higher has increased from 11 in the 1930s to 19 in the late 2010s. In contrast, the number of frost days — days when the temperature drops below 0°C — has decreased by 10 to 5 in the North Island, and from 54 to 40 in the South Island, over the same period. Warm spells (three or more days running with temperatures

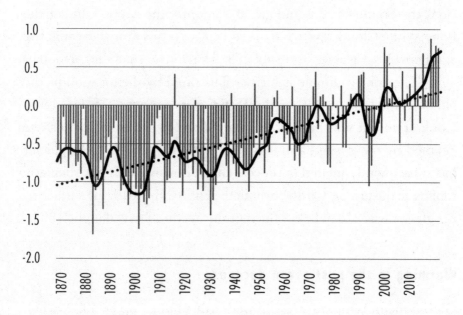

FIGURE 1. New Zealand mean temperatures 1870–2019, compared with the 1981–2010 average. Individual years (bars) are shown as departures, in °C, from the climatology. The smoothed line shows the variations and the dotted line the trend.

above the one-in-ten-day percentage average) have increased and cold spells (the same, but colder than average) have dramatically decreased, the latter from 21 to 4 days a year nationally.

Growing degree days measure the degree of warmth in the surface air temperature which is available for plant and insect growth over a July to June growing season, and can be used to predict when flowers will bloom and crops and insects will mature. Units count the total number of degrees Celsius that each day is above a threshold, normally 10°C. Increased growing degree days means that plants and insects reach maturity faster, provided that other conditions necessary for growth are favourable, such as sufficient moisture and nutrients.

The number of growing degree days per growing season has increased significantly, especially in the North Island. Northern New Zealand, north of Auckland, has been frost-free since 2000, which means that subtropical plants such as bananas will now be able to successfully reach maturity there. This change also means that areas such as the Wairarapa and Marlborough, which have traditionally grown grape varieties that prefer cooler conditions, such as sauvignon blanc, will be able to grow warmth-tolerant grape varieties such as merlot and cabernet sauvignon.

An increase in growing degree days means that in the South Island grain maize can be cropped further southwards in Canterbury, and warmer varieties of wheat can be grown in Southland.

The 'heatwave' summers of 2017/18 and 2018/19

During the summers (December to February) of 2017/18 and 2018/19, there were extreme heatwaves in the seas around New Zealand and over the land area. Heatwaves are events where temperatures are unusually warm for five or more days. These heatwaves ranked as the first and third most intense in the climate record (1934/35 ranked as the second), and covered a huge area of approximately 4 million sq km, which included the south and central Tasman Sea, from just east of Tasmania across New Zealand to the Chatham Islands.

Average air temperatures over the land were 1.7 to 2.1°C above average, while sea-surface temperatures were 1.2 to 1.9°C above average. The above-average summer temperatures for 2017/18 compare best with end-of-century climate warming projections where mean annual temperatures increase by 1.5°C to 2°C nationally over New Zealand. The two heatwaves exhibited the greatest increase in above average recorded temperatures to the west of the South Island.

Many more fine-weather systems — highs or anticyclones — than usual occurred over the Tasman Sea and extended southeast. Stormy weather systems (called lows and troughs) were largely absent. Ice loss in the Southern Alps was estimated at 9 cubic km (km³): 22 per cent of the 2017 ice volume which melted in the combined 2018 and 2019 years.

> **Glaciers are large-scale, highly sensitive climate instruments. Their fluctuations are among the clearest signals of climate change.**

As a result, major species disruption occurred in marine ecosystems, with bull kelp dying along the Canterbury and Otago coast. Farmed fish died in salmon farms in the Marlborough Sounds. Commercial fishers reported that snapper were spawning approximately six weeks early off the South Island coast, and Queensland groper were reported in northern New Zealand, 3000 km out of range.

The Southern Alps glaciers

Glaciers are large-scale and highly sensitive climate instruments. Recent history, since the nineteenth century, has shown general loss of ice from mountain glaciers worldwide. Their fluctuations are among the clearest signals of climate change.

Mountain glaciers are different from the two ice sheets located over Greenland and Antarctica, because they move faster and have younger ice. They are simply created by the surplus of accumulated snow and ice that collects each year above the permanent snowline, where the losses to summer melting are less than the gains from winter growth. This creeping cover of annual layers, the névé, formed over many decades of winter snowfalls, is balanced lower down the mountain by losses to melting. The higher the mountain rises above the permanent snowline, the more surplus snow that collects, and the larger the glacier formed.

A glacier always flows down to cross the permanent snowline from the area of snow gain to the zone of ice loss. This marks the elevation at which, over any one year, snow gain is exactly balanced by melt losses. If the temperature or snowfall changes even a little, the average snowline elevation shifts dramatically, with consequent changes to the gains to and losses of ice from the glacier. Usually, warmer temperatures will raise the snowline and colder temperatures will lower it. Glacier ice amounts are therefore normally related to temperature levels.

Individual glacier-reaction times to changes in climate vary from five to more than a hundred years, so it's risky to use any one patch of ice as a benchmark for regional climate change. Most of the steep alpine glaciers in wet climates respond quickly, rolling a wave of newly gained ice down to their terminus to appear as an advance in five to fifteen years.

The Franz Josef and Fox glaciers, two of the large glaciers in the Southern Alps of New Zealand, are exceptionally volatile. Ice from their snow-gain zone 'squirts' down their steep, narrow valleys in five to eight years, like a tube of toothpaste. They are among the most sensitive of the New Zealand glaciers, and their fluctuating terminus positions give a detailed summary of changes in climate. By contrast, the lower trunks of long, gentle valley glaciers, such as the Tasman Glacier near Aoraki/ Mount Cook, and especially those glaciers carrying massive blankets of rocks which shield their ice from the sun, behave as creeping ponds of ice that have delayed responses to changes in temperature of a hundred years or more. Their response to small 'pulses' of snow gain is to absorb the growth and add their effects over many decades.

Worldwide glacier-length trends are recorded by the World Glacier Monitoring Service (WGMS), based in Switzerland. This inventory provides enough data for calculation of glacier lengths and mass balance, measured in depth of water equivalents. The WGMS has calculated cumulative trends in glacier mass (in vertical metres of water equivalent) in 10 mountain ranges around the world.

Since the late 1950s, apart from three positive mass-balance years in 1965, 1967 and 1968, glaciers globally have thinned at an accelerating rate, because of climate change. Since the 1960s, ice thinning has been dramatic, with about 8 m of thickness lost from 1995 to 2010 alone.

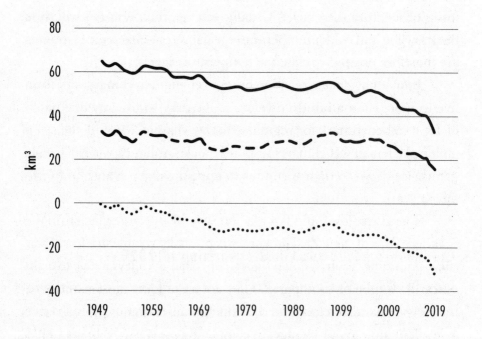

FIGURE 2. Ice-volume changes for the Southern Alps of New Zealand, in cubic kilometres (km³), 1949–2019. The dashed line is the volume loss from the small and medium-sized glaciers, and the solid line includes ice-volume loss from all glaciers. The dotted line is the aggregated volume loss since 1949.

The story from the Southern Alps (shown in Figure 2) is just as remarkable. Using the glacier mass-balance surveys based on glacier snowline elevations since 1977, and Tasman Glacier snowline elevations back to 1949, scientists have found that warming since 1949 has caused much of the ice loss for our Southern Alps glaciers.

For the big, debris-covered valley glaciers, glacial lakes have formed at their fronts. Once a lake has formed, it eats at the glacier far faster than surface melt. At the front of the glacier, the ice cliff calves bergs into the lake. This accelerates a massive and catastrophic depletion of the glacier ice volume, creating an irreversible tipping point for the glacier. It would take an ice-age climate to reverse the process and drive the glacier back across the lake.

As part of his PhD thesis, scientist Andrew Ruddell estimated the Little Ice Age extent of ice volume in the Southern Alps (in 1880) to be 100 km³. Between 1949 and 2019, the volume decreased around a half, from 65 km³ to 32 km³. This loss is an ongoing response to regional warming of about 1°C in mean temperature since the 1890s.

Just under half of this loss — 15 km³ — came from 12 of the largest glaciers. Apart from these large valley glaciers, the loss in ice volume due to annual climatic mass-balance changes from the medium to small glaciers was 17 km³ (see the dashed line in Figure 2). A huge Southern Alps ice loss of 9 km³ occurred during the heatwave summers of 2017/18 and 2018/19.

Greenhouse gases and climate warming to 2020

Since the Industrial Revolution began in the late 1700s, about 380 billion tonnes of carbon have been emitted by humans into the atmosphere as carbon dioxide (CO_2). Atmospheric measurements show that about half of this CO_2 remains in the atmosphere and that, so far, the ocean and terrestrial sinks which absorb atmospheric CO_2 have steadily increased. The Mauna Loa CO_2 record — also known as the Keeling Curve after

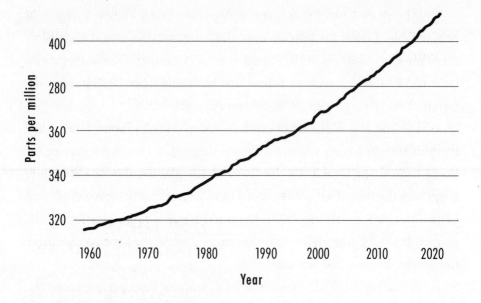

400 —
280 —
360 —
340 —
320 —

Parts per million

1960   1970   1980   1990   2000   2010   2020

Year

FIGURE 3. The Mauna Loa Observatory carbon dioxide (CO2) record, 1958 to 2020.

its creator Dr Charles David Keeling, who began studying atmospheric carbon dioxide in 1956 — is the world's longest unbroken record of atmospheric CO^2 concentrations.

This record, from the National Oceanographic and Atmospheric Administration's Mauna Loa Observatory, near the top of Mauna Loa on the Big Island of Hawai'i at 3397 m altitude, shows that CO^2 has been increasing steadily from values around 317 parts per million (ppm) when Dr Keeling began measurements to over 410 ppm today (Figure 3). Concentrations are currently approaching the symbolically important value of 450 ppm, which represents the level that is required to keep anthropogenic global warming at 2°C or below, to avoid catastrophic climate effects.

Increased levels of carbon dioxide and other greenhouse gasses — methane and chlorofluorocarbons (CFCs) — trap heat in the atmosphere and result in an enhanced greenhouse effect. The basic laws of physics require that an increase in these gases will change climate.

The conclusions of the first report of the Intergovernmental Panel on Climate Change (IPCC), a group of more than 2500 top scientists from around the world, have not changed, despite intensive and continuing research and analysis. Essentially, the increase in greenhouse gases is like a person in bed at night popping on another blanket or two — more of the heat generated by that person gets trapped, which makes the person warmer.

The continued rapid rise in CO_2 ensures that levels in the atmosphere will rise far beyond 400 ppm before they stabilise. If the pace of the past decade continues, CO_2 will reach 450 ppm or more by the year 2040, and this amount will cause 2°C of global warming, the tipping point into catastrophic climate change.

CO_2 is the most significant human-made greenhouse gas,

CO_2 is the most significant human-made greenhouse gas, produced mainly by the burning of fossil fuels such as coal, oil and natural gas.

produced mainly by the burning of fossil fuels such as coal, oil and natural gas. The pace of the rise of CO_2 and other greenhouse gases depends strongly on how much fossil fuel is burned, the rates of deforestation and agriculture methane released by livestock and rice paddies, and possible melting of bubbles of methane in the frozen Arctic.

Between 1900 and 2020, CO_2 increased from 300 to 410 ppm. Brett Mullan at the National Institute of Water and Atmospheric Research (NIWA) has run 13 climate models for the New Zealand area which track the amount of warming that might be expected with this rise in carbon dioxide and other greenhouse gases. The expected temperature increase above average was from 0.1°C in the decade 1900–10 to 0.8°C in the 2010s — a climate warming increase of 0.7°C. In fact, the recorded increase over this period was 0.8°C.

The future

To track how climate will change from now until 2040 and 2090 it is necessary to develop scenarios using different rates and scales of climate change, so as to judge the potential effects on flora and fauna, glaciers, freshwater and New Zealand's surrounding marine resources, livestock and arable and fruit crops, and on health and other impacts on human communities. The IPCC — the primary scientific body that advises on climate change science, its effects, and ways to reduce greenhouse gas emissions — has developed four scenarios. These are called representative concentration pathways or RCPs. Each RCP has a level of estimated emissions of carbon dioxide, methane, nitrous oxide and other greenhouse gases up to 2100, based on assumptions about economic activity, energy sources, population growth and other factors.

SCENARIO 1 — LOWER EMISSIONS SCENARIO: CO^2 concentrations reach a maximum level of slightly above 530 ppm in the year 2080, then decline. In this scenario resurgent nationalism, concerns about competitiveness and security, and regional conflicts push countries to increasingly focus on domestic or, at most, regional issues. Countries focus on achieving energy and food-security goals. Population growth is low in industrialised and high in developing countries.

SCENARIO 2 — HIGH EMISSIONS SCENARIO: The last of the four scenarios has increasing CO^2 concentrations throughout, reaching 950 ppm and rising by 2100 and beyond. This world is business as usual. The push for economic and social development is coupled with the use of abundant fossil-fuel resources and the adoption of high-energy-use lifestyles around the world.

Of the other two scenarios, one is unachievable in keeping temperature increases low, while the other lies between scenarios 1 and 2.

Brett Mullan at NIWA has modelled the mean annual temperature increases for the New Zealand region based on the two scenarios above. The lower emissions scenario 1 shows an increase in mean temperatures nationwide at 2040 of 0.7°C, which is continued in 2100. The high emissions scenario 2 shows a temperature increase at 2040 of 1.0°C, and 3.0°C by 2100. In all cases temperature increases are greatest in summer and autumn, and least in winter and spring.

W ith the pattern of change in average temperature relative to 1995, both the lower and higher emissions scenarios show more warming for inland areas and northern New Zealand, compared with southern New Zealand. Extremes of temperature show larger changes. Under scenario 1, hot days per year — defined as days of 25°C or more — increase by 10 in the west and up to 30 in the north and east. Under scenario 2, the number of hot days per year moves up by 20 days in the west and south of the South Island, and 70 days or more in northern New Zealand.

In contrast, frost days (days of 0°C or below) go down quite considerably. For scenario 1, they decrease by 10 to 20 days nationally. With scenario 2, frost days decrease by 10 to 30 days for much of the North Island, and northern New Zealand is frost free. Frost days go down by 50 to 60 days in the central North Island, and in extensive areas of inland Marlborough, Canterbury, Otago and Southland.

Compared with 1981–2010, rainfall shows a pattern of increases in the west and south of the South Island (Westland, Fiordland, Southland and Otago), and decreases in the north and east (Northland, Auckland, Waikato, Bay of Plenty, Gisborne, Hawke's Bay and Wairarapa). In summer, rainfall decreases are generally accentuated in already dry areas. For winter, the pattern of increases in the west and south is heightened, reflecting the increase in prevailing westerly-quarter winds. The number of snow days decreases substantially in high-altitude and southern regions of the South Island.

Conversely, because a warmer atmosphere can hold more moisture, the potential for more heavy rainfall events is increased. Extremes increase for both heavy rainfall and more dry days.

Almost the entire country shows some increase in heavy rainfall, with increase above 30 per cent in much of Otago and Southland. More extreme rainfall, such as the one-in-100-year events, is expected to go up even more. The number of dry days (with daily precipitation of less than 1 mm) increases over most of New Zealand. The result is a climate of more floods and droughts.

Conclusion

Aotearoa New Zealand's climate journey with planet Earth has experienced large changes over time. In the depths of the last ice age, around 20,000 years ago, average temperatures were about 5°C lower than they are today. It looks now as though global warming is likely to heat up this country and the world between 2 and 3°C by the year 2100, if no reduction in greenhouse gases occurs.

Past climate shifts have had dramatic impacts on the sculpting of the landscape, especially in the South Island, and more recently on human inhabitants since around 1300, particularly with changes in the incidence of floods, droughts, snow- and wind-storms, and other climate extremes. Those differences in average temperature may seem small, but they have a huge effect on the environment and the lives of plants and animals.

Our environment has changed already, and our everyday lives will be affected — from access to food, water and even land, to health and biodiversity — as the world and Aotearoa continues to warm over the next 100 years and beyond. The change in climate forces created by warming will alter the pattern of climate extremes.

The warmer temperatures have already had effects on our

environment, perhaps the most noticeable of which is the marked loss of ice in most of the mountain glaciers of the world. Since 1950, there has been a massive retreat of mountain glaciers. From 1980 to 2016, these glaciers have lost from their surfaces to the oceans an accumulated 22 metres of water equivalent (the depth of water that results if the whole snow-pack melts). Other dramatic impacts have occurred, including especially severe coral bleaching on the Great Barrier Reef. Warming has also pushed up tropical cyclone activity over the western South Pacific, especially near Vanuatu, Fiji and Tonga, and produced severe drought in parts of Africa and Australia.

Progress towards a new global climate agreement has been painfully slow. Unless there is a more coordinated global action to tackle climate change soon, it will be increasingly hard to reduce poverty, in all its dimensions, particularly in the world's poorest countries. The costs of adaptation will also rise steeply everywhere. Fairness is required for the poor, with intergenerational equity for our children and grandchildren.

In December 2015, 195 nations of the world negotiated the Paris Agreement within the United Nations Framework Convention on Climate Change. This represents the opportunity and hope for global greenhouse gas mitigation. Only time will tell if the political commitment is there — but virtually all countries have accepted the negotiated agreement. However, to really work, it still requires ratification and then measures of enforcement.

How will our coasts and estuaries change with sea-level rise? Implications for communities and infrastructure

DR ROB BELL
Managing director, Bell Adapt Ltd
(formerly of NIWA)

I te takutai — along the coast

Ever since early Māori navigated by the stars and godwits (kuaka) to Aotearoa New Zealand, coastal dwellers have had a close connection to the sea. It has provided an abundant food source, a means of transport to skirt around the rugged interior, and stunning landscapes and beaches. Local characteristics such as the tide range and occasional storm and wave effects would have been observed and gradually appraised by early inhabitants, before they put down roots and developed the desirable flat, fertile, coastal margins with access to waterways, in tandem with strategic coastal headlands.

Going back further, mean sea level stabilised around 6000–7000 years ago, having dropped from an early Holocene[1] high-stand sea level of 1–2 m higher than present. It has been more or less consistent over many centuries (Clement et al., 2016; Dougherty & Dickson, 2012).

Early development around our shores assumed that mean sea level was the one constant within the backdrops of highly dynamic processes such as tidal currents, waves, storm surges and river floods, and an active tectonic setting (e.g. tsunamis, earthquake ruptures and volcanoes). All these dynamic processes have shaped our magnificent sandy beaches, cliffs, salt marshes and sheltered harbours.

Extraordinary development of the coast and river mouths occurred soon after World War II, when thousands of families spent weekends and holidays on DIY projects, acquiring a slice of coastal land and putting together the iconic New Zealand bach, to enable them to enjoy some recreational time out. This created beachfront holiday communities and fishing-hut enclaves, such as Selwyn Huts in Canterbury (Peart, 2009). One of my early memories is of staying

---

1   The Holocene is the geological era of the last 11,700 years, since the end of the last major ice age, when sea level was much lower. Since then, there have been small-scale climate fluctuations, but in general the Holocene has been a relatively warm period in between ice ages (an interglacial), when our present beaches and estuaries were established.

at my grandparents' fishing bach at Milford (near Temuka) and going fishing in a dinghy along the tidal reach of the river, looking for salmon or putting out a net to catch 'herring'.

This era of building modest baches with no utility services (other than the gravel access road) was followed from the 1980s by an even bigger development boom, which has continued to the present, of high-value, large, permanent coastal dwellings, resorts and canal estates. These coastal settlements mostly come with full utility services and are accessed by highways that in sections hug the shoreline, such as State Highway 25 up the west coast of the Coromandel Peninsula.

Today, with the benefit of hindsight and tide-gauge monitoring, the premise that mean sea level is constant is no longer valid. Sea level has been rising since around the start of the twentieth century, with a more persistent acceleration in the rate of rise since 1960. It is now around 0.2 m higher around New Zealand than it was in 1900 (MfE, 2017; MfE/StatsNZ, 2019). That doesn't sound like much, but when matched up as a proportion of our storm surges of up to around 1 m — relatively modest compared with countries exposed to tropical cyclones or hurricanes — it's no wonder we are observing more-frequent coastal flooding. Essentially, all the dynamic coastal processes, such as waves, tides and storm surges, and their combination with river floods, are now riding on the back of a higher base mean sea level.

This emergence of a higher base sea level along our coasts and harbours is only the start, as global greenhouse gas emissions to date (even if cut off now) have already committed the world to an average of at least a 1 m rise over the next few centuries. How ongoing emissions pan out over coming decades will determine what accelerated rate of rise in mean sea level eventuates and how large the rise will be before the oceans stabilise again (Oppenheimer et al., 2019).

In the meantime, globally and regionally, citizens and decision-making agencies and councils need to tackle the coastal risk challenge on two fronts: reducing greenhouse gas emissions (mitigation) and managing the impacts and implications of a rising mean sea level that has crept up on us (adaptation).

Coastal adaptation also needs to address other climate-related changes, such as changes in storm intensity, more saltwater up creeks and rivers, rising groundwater levels and the rise in the 'flood sandwich', in which river floods, more intense rainfall and coastal flooding combine to exacerbate flooding. Low-lying coastal settlements around Aotearoa New Zealand are already being adversely affected, such as fishing-hut enclaves, some towns and city suburbs, pastoral land and iwi/ hapū cultural and sacred sites such as marae and urupā.

> **Low-lying coastal settlements around Aotearoa New Zealand are already being adversely affected, such as fishing-hut enclaves, some towns and city suburbs, pastoral land and iwi/ hapū cultural and sacred sites such as marae and urupā.**

Future sea-level rise may follow a range of different possible trajectories, depending primarily on global mitigation efforts and how the polar ice sheets respond to warming, leading to widening uncertainty over time. Adaptation to this ongoing rising risk requires a new paradigm to work adaptively with the uncertain pace of change, summed up by the phrase 'monitor and act pre-emptively' (MfE, 2017; Lawrence et al., 2018, 2020), rather than automatically defaulting to the prevailing mindset of 'defend and hold the line'.

This conventional go-to stance is still prevalent among many beachside-property owners in New Zealand and Australia. For example, after storm erosion on the Central Coast of New South Wales in July 2020, a homeowner responded, 'I get [that] people would say, "Well they shouldn't have built there", but the fact is we have been allowed to build here — we're protecting our homes. This is totally avoidable.'[2] But with the committed rise in sea level likely to linger for several centuries — the rate of rise depending on the effectiveness of reducing global emissions — holding the line is becoming unsustainable.

Maybe the words 'shoreline' and 'coastline' are really misnomers. Natural variability has seen shores move to and fro over recent history, responding to storms and quieter beach-building periods. However, some areas in Aotearoa New Zealand have become coastal erosion hotspots. Ongoing sea-level rise will manifest in many more areas as a gradual movement of these shores inland, on top of the natural fluctuations in erosion and accretion. For coastal cliffs, it is all one way — landward retreat.

This chapter sets out why sea level is rising and how areas along our coast and upstream into lowland rivers will change with sea-level rise and climate change. What follows are the implications for low-lying communities, iwi/hapū and infrastructure, and how best to work adaptively around the impending changes, rather than simply defaulting to 'holding the line' and hunkering down (for a while anyway).

There are other short-term actions and long-term options, but there is no denying that adaptation to the ongoing rise in sea level will be complex and heart-wrenching for coastal dwellers who have a strong attachment to their homes and whenua. However, New Zealanders are not in this alone — it is a global risk facing around 630 million people who currently live on land below annual coastal flood-levels projected for 2100 (Kulp & Strauss, 2019). While we will not see the gravity of the emerging risks facing deltas and islands such as Bangladesh and the Mekong Delta in South East Asia and some Pacific Islands, nevertheless around 178,000 coastal residents in Aotearoa New Zealand will potentially be exposed to extreme coastal-storm flooding on top of a 1 m rise of sea level (Paulik et al., 2020).

---

2   https://www.msn.com/en-au/news/australia/central-coast-houses-in-danger-of-collapse-as-large-swell-causes-unprecedented-coastal-erosion/ar-BB16PLQM?ocid=msedgntp

The sea is a-changin', and it's all one way

After being relatively stable for millennia, mean sea level started to rise by around 1900 in Aotearoa New Zealand, based on salt-marsh borehole analyses (Gehrels et al., 2008; King et al., 2020). While there are few, the longest available historical tide-gauge records globally, extending back to around 1800 in the northern hemisphere (e.g., Amsterdam, Liverpool, Stockholm and Brest), show that ocean levels started rising from the late 1800s onwards (Woodworth et al., 2011).

New Zealand's longest-standing tide-gauge records from the four main ports,[3] starting around 1900, show an underlying steady but sure rise of 0.18–0.2 m in mean sea level in our waters, relative to the landmass the gauges sit on (Figure 1).

This underlying upwards trend sits behind year-to-year variability produced by climate cycles such as the two- to four-year El Niño–Southern Oscillation and the longer, 20–30-year Interdecadal Pacific Oscillation.[4] As an example, during La Niña events, annual mean sea level is often higher than normal (e.g., years 1987–88, 1999 in Figure 1), while it drops somewhat during El Niño events (1993, 2010). Two of the projections (high and low) for Aotearoa New Zealand out to 2030 are also plotted (dashed lines) in Figure 1 to show how the annual mean sea level is currently tracking in comparison to modelled projections for rises in sea level. Initially, the range of future projections is narrow, because of the delayed response of seas rising, but it will spread further apart towards the second half of this century and beyond, depending on the effectiveness in reducing global emissions.

From 1900 up to 2018, the average rise in mean sea level in our waters has been 1.77 mm per year [±0.05 mm] (MfE/StatsNZ, 2019),

---

3   Dunedin Wharf, Lyttelton, Wellington, Auckland.
4   Also related to the Pacific Decadal Oscillation, which is a longer-term, Pacific-wide climate
      cycle that alternately emphasises more La Niña–like (negative phase) or more El Niño–like
      (positive phase) climate settings.

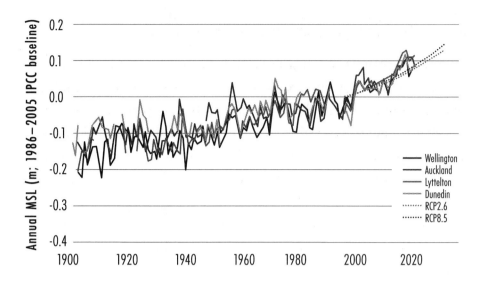

FIGURE 1. Annual mean sea level (MSL) from New Zealand's four long-term gauge records, up to the end of 2019. MSL is relative to the gauge-record average over the period 1986–2005, which is the baseline (zero) used internationally for future projections. (Data from port companies [Auckland, Wellington, Lyttelton, Dunedin] compiled by R. Bell.)

which is slightly higher than the global-average rate of 1.56 mm/yr [±0.32] over the same period (Frederikse et al., 2020).

Splitting the long New Zealand tide-gauge records into approximately 60-year halves reveals that the rate of rise has accelerated two-fold in recent decades, with an average rise of only 1.21 mm/yr [±0.11] from 1900 to 1960 increasing to 2.44 mm/yr [±0.10] from 1961 to 2018 (MfE/StatsNZ, 2019). More recently, since 1993, called the 'satellite era', radar altimeters on a sequence of satellites have been regularly orbiting Earth measuring the height of the sea surface along preset tracks between 66° north and south. These more extensive measurements over the vast oceans, albeit over a shorter period, show that the rise has increased further to a rate of 3.5 mm/yr up to July 2020.[5]

The close alignment of New Zealand measurements with

_____

5   https://www.cmar.csiro.au/sealevel/sl_hist_last_decades.html

HOW WILL OUR COASTS AND ESTUARIES CHANGE WITH SEA-LEVEL RISE?

global-average trends in sea-level rise confirms that use of future global projections of sea-level rise are generally applicable to us here, noting there are known regional variations in different parts of the world's oceans and seas. For the wider Southwest Pacific regional sea centred around New Zealand, the modelling shows we can expect a small additional rise by 2100 of up to 5 cm more than the global average projections (Ackerley et al., 2013). More on the future prospects later.

What is causing mean sea level to rise globally?

The Earth's atmosphere is warming. Averaged globally, present-day surface temperatures have risen by around 1.0°C since 1880.[6] Human activities are estimated, with very high confidence, to have caused the warming, from global emissions of greenhouse gases (IPCC, 2019a). In particular, long-lived atmospheric carbon dioxide is the most relevant greenhouse gas fuelling the rise in sea level in the long run.

Several factors have contributed to global, regional or local rise in sea level in the recent past and are likely to have an impact well into the foreseeable future (MfE, 2017; Oppenheimer et al., 2019; Frederikse et al., 2020). Different response times apply to the three main contributors:

1. **THE VOLUME OF THE OCEAN**, which has increased from thermal expansion of ocean waters, mostly in the top 2000 m depth. As the ocean warms, the water density decreases, thereby increasing the volume. With the exception of low-lying areas, most of the world's shorelines constrain the sea moving too far inland, so most of the volume expansion of the oceans is expressed as a rise in the surface sea level. This is one of the major contributors to sea-level change during the twentieth century.

---

6   https://www.giss.nasa.gov/research/news/20190523/

2. **WATER MASS** added to the oceans from the melting or break-up of land-based ice stores such as glaciers, ice caps and polar ice sheets (particularly in Greenland and West Antarctica). The contribution of additional water mass is accelerating and in the long term will be the largest component of sea-level rise, as ice sheets diminish.

3. **VERTICAL LAND MOVEMENT** (uplift or subsidence), which can significantly alter ocean sea-level rise locally. Especially important is ongoing subsidence, which exacerbates the sea-level rise such areas will experience.

Other contributors to the rise include changes in net storage of water on land (e.g., groundwater extraction), changes in ocean currents, and changes in the Earth's crust as ice mass is lost from the polar ice sheets.

Sea-level rise will always be one of the last environmental responders to warming, due to the long lags for each incremental change in air temperature to work its way through the vast connected oceans and ice stores on Earth. Any surface warming will take many decades to mix through the deep, extensive oceans, which contain 97 per cent of the Earth's water, with a huge volume of 362 billion km^3 and an average depth of 3700 km.[7] Further, the substantial polar ice sheets in Antarctica[8] (which incidentally hold around 60 per cent of the Earth's freshwater) and Greenland will take even longer to respond to both atmospheric and ocean heating. Therefore, these delayed responses mean continued global sea-level rise lingering over several centuries. Importantly, though, reduction in emissions is still critical, as it determines both the short-to-medium-term rate at which seas continue to rise and also ultimately what overall rise in sea level is reached once the climate–ocean system is stabilised.

---

7   https://www.ngdc.noaa.gov/mgg/global/etopo1_ocean_volumes.html.

8   The Greenland and Antarctica ice sheets store the equivalent of 7 m and 58 m of global sea-level rise respectively [Oppenheimer et al. 2019], but this would be realised only over millennia, with no effective reduction in emissions in the foreseeable future.

What is the expected future rise in sea level?

There is high confidence from a recent Intergovernmental Panel on Climate Change (IPCC) assessment that global mean sea level will continue to rise for centuries, well beyond 2100, due to ongoing deep-ocean heat uptake and mass loss from polar ice sheets [IPCC, 2019a; Oppenheimer et al., 2019]. The amount of longer-term sea-level rise depends on future emissions. IPCC mainly uses four emission projections to represent different climate futures, called Representative Concentration Pathways (RCPs). The lowest RCP 2.6 involves rapid reductions in emissions to a net-zero level by around 2075, whereas RCP8.5 represents continuing high emissions with no effective global reduction. The mid-range global sea-level rise for a high-emissions future (RCP8.5) is 0.84 m by 2100, which is around double the rise (0.43 m) if emissions were to follow the low-emissions pathway (RCP2.6) instead (IPCC, 2019b).

Presently, national guidance from the Ministry for the Environment (MfE, 2017) provides for four scenarios of future sea-level rise in Aotearoa New Zealand (Figure 2). Use of all four scenarios is recommended to ensure that future planning and adaptation options are sufficiently flexible to cope across a range of possible coastal futures. It is not clear at present, from sea-level measurements at long-term tide gauges to date, which of the four scenarios the rise will track along (or it could be a combination over time). For new development and adapting of existing assets and infrastructure, we need to stress-test all future plans, decisions and project designs against all four scenarios to ensure they are flexible and robust enough, to avoid locking in unsustainable responses down the track for future generations. Nor do we want to over-design our initial responses to match the worst case — rather the key is building in flexibility to adjust later, with sufficient lead time.

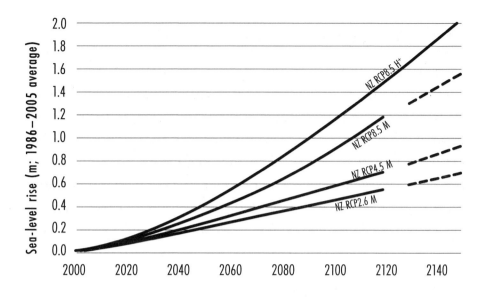

FIGURE 2. Four scenarios for New Zealand-wide sea-level rise projections recommended in the Ministry for the Environment's 2017 coastal guidance, based on the IPCC 5th Assessment Report (Church et al., 2013), out to 2120, with extensions to 2150 and the highest scenario based on Kopp et al. (2014).

Coastal hazards on the rise

Keen observers at coastal and harbour areas will have noticed over time the increasing occurrence of many harbours and estuaries being brim-full during days with king (perigean-spring)[9] high tides, even when it is sunny with no storm in sight (hence the term 'sunny-day flooding'). This increasing frequency of harbours 'bursting their seams' is the first sign of the ongoing rise in the mean sea level. That is because tides, waves, storms and tsunamis ride on the back of that base mean sea level — essentially ratcheting upwards any dynamic coastal process. While a rise of 0.3–0.4 m in mean sea level doesn't seem that much (e.g., calf-muscle height), the initial impact will manifest as more frequent

---

9   King or perigean-spring tides peak every seven months, when the new or full moon closely coincides with the moon at its closest to the Earth in its ~27-day elliptical orbit (called the moon's perigee).

nuisance and damaging coastal flooding during storms, as water levels are pushed up by storm surges and/or when swells are present (such as on the southern Hawke's Bay coast).

Coastal flooding will become more commonplace over the next few decades. In New Zealand we are not affected by direct hits from tropical cyclones or hurricanes, unlike other parts of the world where storm surge can reach several metres. However, although our present-day extreme storm tide[10] events are typically only another 0.35–0.45 m higher than a normal king high tide, only modest rises in sea level of a similar height will change flooding from a rare infrequent event (in the recent past) to something that occurs once or more a year on average (PCE, 2015). In areas with small tide ranges, such as Wellington (Table 1), rises in relative sea level of only around 0.3 m will mean that previously rare coastal flood events will turn up on average once a year, and such a sea-level rise

Sea-level rise will also exacerbate shorelines known to already be hotspots for coastal erosion

is expected by around 2040–50. Auckland, which has a greater tide range, would require a sea rise of 0.45 m for a similar change to an annual occurrence.

Sea-level rise will also exacerbate problems along shorelines known to already be hotspots for coastal erosion, but unlike coastal flooding, the general climate change influence on erosion of sandy/gravel shores and cliffs around our coast is less predictable. Erosion or accretion of shorelines typically exhibits variability over years to decades owing to complex interactions arising from climate cycles and wave patterns, sediment abrasion rates, the delivery of sediment to the coast

---

10   Storm tide is the heightened sea level at the shore during storms combining a high tide, storm surge (when the barometric pressure drops and/or onshore winds pile water up against the coast), wave set-up and the state of mean sea level (e.g. in the late summer, when it is warmer, and during La Niña conditions, the background sea level is usually higher than normal by several centimetres).

Sea-level rise (cm) needed for a past 100-year flooding event to occur . . .	Average occurrence of extreme coastal flooding for the 4 main centres			
	AUCKLAND	WELLINGTON	CHRISTCHURCH	DUNEDIN
0 cm	Every 100 years	Every 100 years	Every 100 years	Every 100 years
10 cm	Every 35 years	Every 20 years	Every 22 years	Every 29 years
20 cm	Every 12 years	Every 4 years	Every 5 years	Every 9 years
30 cm	Every 4 years	Once a year	Once a year	Every 2 years
40 cm	Every 2 years	Every 2 months	Every 3 months	Every 9 months
50 cm	Every 6 months	Twice a month	Twice a month	Every 3 months
60 cm	Every 2 months	Three times a week	Twice a week	Once a month
70 cm	Every month	Every tide	Every day	Once a week
80 cm	Every week	Every tide	Every tide	4 times a week
90 cm	Twice a week	Every tide	Every tide	Every tide
100 cm	Every day	Every tide	Every tide	Every tide

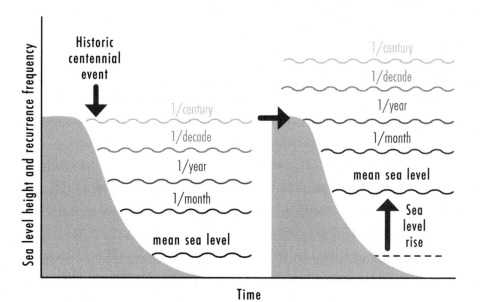

TABLE 1. Occurrences of extreme coastal flooding events in the four main coastal centres will increase from what used to be a rare 100-year (centennial) event to increasingly occur more often as the mean sea level continues to rise. The table shows how the ongoing sea-level rise (SLR) will reduce the average period (from 100 year in the recent past) for such extreme events to occur in 4 cities. The schematic shows an example of how this change from a rare extreme event in the recent past will occur once a year on average due to a higher sea level (e.g., in Wellington this will occur when SLR reaches 30 cm shown in the table) [After PCE(2015) and Fig SPM.4 of IPCC(2019b)].

HOW WILL OUR COASTS AND ESTUARIES CHANGE WITH SEA-LEVEL RISE?

and from offshore, and the degree of intervention by shore protection, beach nourishment or dredging. Therefore it is not always clear-cut how climate change and sea-level rise will specifically affect shorelines in each locality. However, the general tendency will be for increasing issues with erosion as sea level rises. The erosional trend is clearer for pocket beaches (sand surrounded by a rocky backshore and headlands) with little ongoing input of sediments to become drowned over time as seas rise, and coastal sedimentary cliffs will tend to erode more quickly than in the past.

Climate change is also expected to affect and in most cases increase waves and storm surges, through increased winds and more-intense storms. This will add to the coastal-flooding mix, but the changes envisaged for New Zealand from climate–ocean models indicate that these effects will be minor in comparison to sea-level rise. For instance, changes to storm surges by the latter part of this century may range from slightly lower than present to slightly higher (Cagigal et al., 2020). Seasonal cycles in mean sea level and variability from El Niño–Southern Oscillation are also likely to increase seasonal and monthly peaks in mean sea level from ocean warming (Widlansky et al., 2020), but again they are still secondary to the main event, which is sea-level rise.

The emerging climate risk affecting coastal and freshwater lowland areas will arise from combinations of hazards (Rouse et al., 2017) — factors that are individually affected by climate change and/or sea-level rise but together lead to worsening flood situations. For example, rising groundwater, combined with river floods, more-intense rainfall and a coincident storm-tide, all elevated by a higher base mean sea level, will lead to increasing frequency of the so-called 'flood sandwich' for coastal lowlands. This will cause ongoing issues with drainage, stormwater management, land use (e.g., the type of agriculture or horticulture being undertaken, although it may also provide other opportunities) and increased exposure to liquefaction during earthquakes. As an example, current reliance on gravity systems to drain most coastal areas will be increasingly compromised and may require adaptation to expensive pumped or vacuum systems similar to that used in Napier, which relies on a pumped stormwater network for three-quarters of the city.[11]

An additional emerging potential hazard from rising sea level is increased salinity of lowland freshwater systems such as streams, rivers, groundwater aquifers, wetlands and coastal lakes, as seawater penetrates further inland, or through wave and tidal overtopping of banks.

Impacts and implications of rising seas

'Every centimetre of sea level rise globally means about a million people will be displaced from their low-lying homelands,' according to Professor Andy Shepherd, director of Leeds University's Centre for Polar Observation and Modelling.[12] Around 250 million people globally live on low-lying land no higher than 1 m above current high tide (Kulp & Strauss, 2019), so coastal communities worldwide must prepare themselves for a much more challenging future, where the rise in mean sea level is accelerating, leading to more frequent coastal flooding and, in many locations, exacerbating coastal erosion.

While Aotearoa New Zealand does not have a large proportion of its residents exposed to coastal flooding on the back of sea-level rise over the rest of this century, nevertheless a significant number will be adversely affected in both individual households and communities in low-lying coastal areas. A recent coastal risk-exposure assessment undertaken nationally by NIWA for the Deep South National Science Challenge (Paulik et al., 2019; 2020) shows that nearly 178,000 (approximately 4 per cent) of the normally resident population in New Zealand are potentially exposed to extreme coastal-flood events on the back of a sea-level rise up to 1 m, increasing to 281,000 people exposed

---

11   https://www.napier.govt.nz/services/water-2/stormwater/our-stormwater-system/
12   https://www.theguardian.com/environment/2020/aug/23/earth-lost-28-trillion-tonnes-ice-30-years-global-warming

up to a 2 m rise (Table 2). This high-level national study assumes flood exposure could occur either directly, from overland seawater flooding, or indirectly, posing a residual risk to land and assets behind seawalls, stop banks or tide gates, through rising groundwater or in combination with river floods — in either case, the primary exposure arises from the land being *below* coastal flood-water levels. In terms of assets and infrastructure, there is a substantial exposure for buildings of all types, roads and railways in low-lying coastal areas, which together with the impact of utility services on residents and communities poses a challenge in adapting to this degree of rising sea level. The uncertainty is how soon this exposure occurs (i.e., the rate at which 1 m of sea-level rise is reached), which could be as early as 2090, or much later, towards 2200.

Clearly, global effort to reduce emissions will have a profound effect on the impacts on low-lying coastal areas worldwide. Focusing on what a global emissions reduction would mean for New Zealand coastal areas, an extreme coastal flood with a sea rise of 0.9 m by 2100, representing the high-emissions scenario, compared with a rise of 0.5 m in the low-emissions scenario, would mean (based on Paulik et al., 2020)[13]:

1. an extra 32,400 buildings affected in coastal areas;
2. an additional $11.6 billion (2016 value) in building replacement cost;
3. a further 46,000 residents (2013 Census) exposed to extreme flooding; and
4. an additional 600 km of roads at risk.

These figures represent the extra adaptation effort which will be required in New Zealand up to 2100, if global emissions are not substantially reduced in coming decades to a low-emissions (RCP2.6) pathway. These figures do not include the substantial extra adaptation that would be

---

13   https://www.mdpi.com/2071-1050/12/4/1513

required beyond 2100 if net global emissions are not reduced to zero by then.

Therefore, long-term we need to work with both sides of the climate change 'coin'. On one side, we need to plan for local adaptation to the delayed sea-level rise we are already committed to from global emissions to date (and beyond); and on the other side, improve our contribution to global mitigation (reduction) of greenhouse gases to put the brakes on the accelerating rise in ocean levels globally. A useful, concise phrase is that adaptation is managing the unavoidable, and mitigation is avoiding the unmanageable (Ginzburg, 2007), which is very pertinent to the long-run sea-level rise situation globally.

Coastal-flooding risk exposure (NZ)	0–1 m sea-level rise	1–2 m sea-level rise
Residents (2013 Census)	177,600	+ 103,600
No. of buildings (all types)	125,600	+ 70,490
Replacement cost of buildings (2016 $NZ)	$38 billion	+ $26 billion
Road length exposed (km)	2856	+ 1376

TABLE 2. The exposure of people, buildings and roads in coastal areas for the first metre of sea-level rise, followed by a lower increase in exposure following the second metre of sea-level rise.

Consequently, low-lying coastal communities, buildings, facilities and cultural sites of iwi/hapū/whānau, and coastal infrastructure such as roads and railways and stormwater and drainage networks, will increasingly be affected this century as coastal flooding becomes commonplace.

Further research is under way through National Science Challenges to augment the above assessments, nationally and regionally. This includes understanding the exposure in coastal areas of Māori whenua holdings, cultural facilities and sites (e.g. marae, urupā) and, after a national-scale stocktake of coastal and cliff erosion, assessing the exposure of people and buildings facing future erosion impacts to complement the coastal flooding assessment discussed above.

A key conclusion from these national-scale assessments is that much of the coastal exposure of people and assets in Aotearoa New Zealand will occur in the first metre of sea-level rise, before it tails off for higher rises in sea level. This is because New Zealand is geologically young and still being shaped by dynamic tectonic processes, so the previously sought-after but limited flat land around harbours, estuaries and coasts has already been developed, before the land rises into more hilly landscapes in many situations, e.g. Nelson, Wellington/Lower Hutt, Napier, Auckland. This means the adaptation challenge for coastal areas will need to be focused on the short-to-medium term, rather than thinking that the risks from rising sea level are some way down the track.

Adaptation to the emerging and ongoing risks

In relation to rising seas, the exposure of tangible assets to more-frequent coastal hazards is only part of the risk faced by coastal communities and hapū/whānau. Sustainable adaptation also needs to address vulnerability to the growing impacts that will cascade and interconnect across the four elements in Treasury New Zealand's Living Standards Framework (social, natural environment, financial and built environment, and human).[14] In this context, vulnerability is often defined as a mix of the following two factors:

1. How sensitive are the systems or coastal communities to climate change and hazard exposure?
2. Is there sufficient coping and adaptive capacity present in the

---

14   Using the Living Standards Framework defined by NZ Treasury: https://treasury.govt.nz/information-and-services/nz-economy/higher-living-standards/our-living-standards-framework

community, business (including Māori businesses and holdings) or the governance systems to achieve sustainable adaptation that reduces or avoids the emerging risks?

Adaptive capacity covers a range of responses and behaviours to change, our perceptions of risk, our level of social and cultural cohesion, the capacity and responses of our governance systems (e.g. local and central government, co-management with Treaty partners and infrastructure providers) to pre-emptively manage changing risk, and finally the level of present and future equity issues and access to funding for adaptation. As an example, Māori have demonstrated their adaptive capacity to change since arrival in Aotearoa New Zealand (Hikuroa, 2020); climate change and sea-level rise will again allow them to demonstrate their adaptiveness, through their shared principles or mātāpono from a te ao Māori perspective. However, their adaptive capacity could be constrained by access to funding arrangements to implement adaptation of their holdings and cultural sites exposed to rising seas.

Key issues around adaptive capacity currently being canvassed nationally are:

1. Are our institutional, statutory and governance systems nimble enough to embrace more-adaptive decision-making? (For instance, the RMA reform panel[15] suggests that a more fit-for-purpose set of planning acts is needed.)
2. What sustainable funding models and principles are best for tackling coastal adaptation (both engagement and implementation)? This could be a mix of targeted and general rates, taxes or levies set aside in a national fund, as suggested by several parties, e.g. Boston and Lawrence (2018); MfE and Hawke's Bay Regional Council (2020); LGNZ (2020).

---

15   https://www.mfe.govt.nz/rmreview

Further still, for Māori communities dealing with climate change impacts and risks, the importance of leveraging economic support and technological resource pathways has been identified. Other key aspects are strengthening sociocultural networks and related cultural conventions and values, learning new strategies and practices, and integrating climate change into iwi/hapū management planning (D.-N. King in Rouse et al., 2017).

Given the potential exposure to rising seas, nationally and at the local level we can and need to start engagement with potentially affected parties now, so they can be ready to plan for disruption from more frequent events and the cumulative financial (insurance, banks, clean-up costs), business, cultural and social impacts. The importance of the emerging coastal risks for our nation is emphasised in New Zealand's first national climate change risk assessment, released in August 2020, where of the top ten priority risks identified, nine are directly relevant to coastal climate change impacts, across five domains (human, governance, economy, natural environment and built environment). These rising risks need to be addressed urgently in the forthcoming first National Adaptation Plan being developed by government (MfE, 2020).[16] A number of councils have also commenced engagement with potentially affected communities, being guided by both a realisation that recent flooding events experienced around New Zealand (e.g. the coastal flood events of January–March 2018, and the Napier floods in November 2020) are a precursor to more-frequent events, and the release of official coastal hazards and climate change guidance in 2017 (MfE, 2017).

When taking in the challenge ahead for our coastal areas, and the widening uncertainty of what sea-level rise could transpire by the end of this century and beyond, it initially appears a daunting task to plan ahead

---

16   A national climate risk assessment and a National Adaptation Plan are required to be produced at no greater than six-yearly intervals under the Climate Change Response (Zero Carbon) Amendment Act 2019 [Part 1C].

with any certainty. When is the right time to implement options to adapt to sea-level rise, erosion and the flood sandwich? Fortunately, a growing body of work internationally specifically addresses how decision-making can be undertaken that squarely faces complex problems which exhibit widening uncertainty (Oppenheimer et al., 2019). This is particularly pertinent to coastal situations worldwide facing rising sea level — but without knowing by how much and when.

Such adaptive frameworks or tools are called Decision-Analysis Methods (or Decision Making under Deep Uncertainty), described in more detail by Marchau et al. (2019). They shift from the conventional approach of analysing the projections and choosing the most-likely or worst-case scenario and implementing the 'best solution' for the chosen scenario (Rouse et al., 2017). This front-loads investment or plan changes (which may not be needed for a while), and could close off flexibility to change tack if further adaptation is required in the future. Instead, these adaptive decision-analysis tools or frameworks address the need to make decisions or plans that are flexible and agile, to enable a switch to more-sustainable alternatives, and that cover adaptation needs across a range of possible future conditions or rates of change. These approaches go hand-in-hand with monitoring the changes over time to inform timely implementation of the next option — not too soon and not too late (to avert undue flood or erosion damage).

The Ministry for the Environment's 2017 coastal guidance leverages this body of knowledge, built around a 10-step decision cycle for coastal adaptation (Figure 3). This is framed around one of these tools, called Dynamic Adaptive Pathways Planning. DAPP provides a framework with community, iwi/hapū and stakeholder engagement at the centre, working together to map out sustainable adaptation options (pathways) for both the coming decades and future generations. This is appropriate given that seas will continue to rise for centuries, and also that New Zealand's statutory processes require planning for 'at least 100

FIGURE 3. The 10-step decision cycle for coastal adaptation, based around five key questions. (Adapted from MfE [2017] coastal guidance. Graphic: after Max Oulton

years' for coastal environments.[17] It is important that potentially affected communities and iwi/hapū gain a broad understanding of the rising coastal impacts and what the 'enough is enough' adaptation threshold is for a specific locality before jumping straight to solutions (as we often do), by diving into a decision-making process. For instance, Step 1 of the decision cycle requires significant groundwork, as it is new territory for all parties to get past the notion of a stable sea level in the past, with changing risk the 'new norm'.

Given the ongoing, changing risks around our low-lying and cliffed coastal areas, community engagement can no longer be a one-off collaboration or consultation process — rather, it is more like a journey to continue planning well ahead of further changes, acknowledging that ongoing engagement needs to be recognised as essential for people's wellbeing and resourced properly (LGNZ, 2020; Stephenson et al., 2020).

---

17   The 100+-year timeframe for planning in coastal environments is prescribed in Policies 10, 24, 25, 27 of the New Zealand Coastal Policy Statement, which is a mandatory national policy statement under the umbrella of the RMA.

Through a collaborative engagement process, DAPP produces an overview of alternative pathways (made up of a mix of short-term actions and/or long-term options) that can be kept open into the future, and when to switch to the next in the sequence or pathway as the adaptation threshold approaches locally. This is usually compiled in the form of a succinct metro-style map. These pathway maps can also include indicators to monitor that provide:

1. **EARLY SIGNALS**, such as reducing levels of service from council infrastructure (e.g. more outages or disruption for sewerage, stormwater, roads or stop banks) or the start of financial tightening (e.g. insurance and bank mortgages being harder to obtain), in tandem with
2. **TRIGGERS** (decision points), where the next option in an adaptive pathway is agreed on and implemented, with sufficient lead-time to avoid the adaptation threshold (MfE, 2017; Stephens et al., 2018; Haasnoot et al., 2018).

DAPP also provides insights into the shelf life (use-by date) of different options and how flexible they are to being upgraded. For instance, enabling a short-term action to buy some breathing space, such as beach nourishment, may be sustainable for a decade or so, and keeps options open (e.g. not physically difficult to stop or remove, unlike a seawall) for changing to the next alternative in a pathway, which could be a form of managed retreat. In a nutshell, DAPP asks the questions:

1. Under what conditions does the plan, design or option start to fail or is no longer tolerable (the adaptation threshold)?
2. What are the alternative pathways or options that enable objectives, levels of council utility services and community and iwi/hapū core values to be achieved over time in a rising risk environment?

DAPP provides a framework for councils collaborating with communities to explore a range of different actions and options to join in various combinations to make up pathways to adapt to rising seas and compounding flooding issues. Generally, options can be categorised into three main types, besides 'do nothing' (Bell et al., 2001; Rouse et al., 2017; Oppenheimer et al., 2019):

1. **ADAPT OR ACCOMMODATE:** such as raising buildings; dune nourishment and planting; limits on property development and intensification; planning incentives to develop on higher ground (e.g. Mapua/ Ruby Bay in Tasman District); creating artificial islands for habitat.

2. **MANAGED RETREAT:** moving dwellings further back on coastal sections; rerouting roads further inland; eventually shifting at-risk parts of a community to an alternative, higher location; realignment of estuary shorelines inland (by deliberate removal of stop banks or walls) to allow wetlands and marshes to migrate at higher tides. In most very low-lying coastal and harbour settings around New Zealand, some form of pre-emptive managed retreat will be inevitable in most situations, other than perhaps in urban CBD areas. It is a matter of *when* it might be required, depending on the exposure and vulnerability of the area to rising seas. However, clearly such a move needs substantial ongoing engagement with residents, to ensure the retreat occurs after a lengthy lead time, which could be decades if not an imminent issue (Lawrence et al., 2020).

3. **PROTECT:** either hard-engineered structures (e.g. seawalls, stop banks, rock revetments, groynes, reclamation) or soft-engineered nourishment of beaches by bringing in external supplies of gravel or sand to match the existing sediment characteristics (e.g. Westshore in Napier and Oriental Bay, Wellington). Protection measures are usually designed to 'hold or advance the line' against

the sea or harbour waters, but can incur significant downsides, such as erosion at the ends of the structure, loss of intertidal beaches, constrained public access to the shore, and the hidden spectre of substantial residual risk from damage, breaching or overtopping — seldom appraised due to an underlying perception of being 'safe' behind such structures, expecting that they will withstand all extreme conditions. Examples of such failures include the Edgecumbe flood stop-bank breach in April 2017, and wave overtopping in Ōwhiro Bay, Wellington[18] in April 2020).

In undertaking future planning to adapt, we need to avoiding locking in inflexible short-term fixes (e.g. hunkering down behind seawalls, which in many situations would need to be topped up and widened

In undertaking future planning to adapt, we need to avoid locking in inflexible short-term fixes.

regularly at considerable expense). Rather, we need to provide for an enduring suite of options (pathways) to meet the foreseeable needs of future generations — which is a purpose that is fundamental to the RMA (section 5) and one of the core principles or mātāpono from a te ao Māori perspective of kaitiakitanga.

'Serious games'[19] are increasingly being used as a tool to engage people to look at different angles of the impending adaptation challenge, especially if they haven't yet experienced coastal-hazard events like flooding. They let participants experience how to make decisions in a safe environment, with constrained financial resources and limited foreknowledge, in response to uncertain changes in climate, economic, social and cultural scenarios that come their way during the game simulation. Participants have to work out when best

---

18   https://wellington.govt.nz/your-council/news/2020/04/big-waves-hit-wellington-coast;
      https://www.youtube.com/watch?v=JXG5b9UgiqU
19   https://niwa.co.nz/natural-hazards/our-services/serious-games-as-a-tool-to-engage-people

to respond and adapt, by simulating how situations might play out for future occurrences of damaging hazard events, followed by a debriefing session on what they learnt and how decision making could be better handled.

Conclusions

Unfortunately, rising sea levels will be with us for a very long time (centuries). Global emissions reductions will make a substantial contribution to both arresting the rate the oceans rise over time (a slower rise allows for more-considered responses) and the ultimate rise, when the climate-ocean-ice system finally reaches equilibrium. From the long-term projections out to 2300 of global sea-level rise at the low and high end of emissions, the ultimate rise in ocean level over centuries could range from around 1 m up to several metres if high emissions persist. So global emission reduction does matter enormously, especially in the long run and for slowing down the pace of change that will be required to adapt to coastal lowlands.

Despite the adaptation challenges and uncertainties for low-lying coastal and estuarine areas facing this ongoing sea-level rise, there are adaptive decision-making approaches such as DAPP or other adaptive decision-making frameworks and tools that enable adaptation to be sequenced, instead of trying to solve the problem once, upfront, by choosing a particular future sea-level rise for the local area. These adaptive approaches enable adaptation planning to be collaborative, creative and pre-emptive, instead of simply reacting to increasingly damaging flood or erosion events and the gradual alteration of coastal lowland environments to more saline and saturated conditions. Further, they keep alternative pathways open, rather than second-guessing the future rate of sea rise over a 100-year planning window and taking a single investment perspective based on a most-likely or worst-case scenario. It is best to get the initial options and actions implemented, as long as they provide future flexibility or a flexible exit strategy, along with putting the brakes on further development in low-lying areas.

Adaptive approaches do, however, rely heavily on monitoring indicators of changes (early signals) until a decision point (or trigger) is reached, when switching to the next option or another pathway needs to be implemented. If sea-level rise slows down, then the planning for the next option can proceed at a slower pace; or, if mitigation targets on global emissions are not met and ocean levels rise faster than anticipated, then implementation of the next plan can be brought forward. However, in a number of low-lying locations around Aotearoa New Zealand, managed retreat will eventually be needed, so further research and policy development is underway to enable, rather than frustrate, its implementation through our governance and statutory processes that are also equitable and fair.

Essentially, coastal areas will experience ongoing gradual change, interspersed with more-frequent hazard events, but that change cannot be managed if it is not monitored. We can gear up to better enable adaptive approaches in a revamped statutory framework, which deliberately addresses rising climate risks and uncertainty, and also provide adequate and equitable funding and resources for adaptation. In tandem, we must also change our fiscal reluctance towards monitoring so that we stay in front of the changes and shifts in the natural environment that can cascade or spread through the human/cultural, economic/financial and built environment domains.

The days of a stable mean sea level are now consigned to history, but the methods are available to adaptively plan our way forward to keep ahead of the sea-level curve.

ABSOLUTE AND LOCAL SEA-LEVEL RISE

Absolute or eustatic sea-level rise (SLR) is the rise in the average ocean-surface elevation measured from the centre of the Earth, usually by satellites with radar altimeters orbiting Earth along preset tracks. Absolute SLR is also the variable calculated by global climate models for future projections and Intergovernmental Panel on Climate Change (IPCC) assessments.

Local (relative) SLR is the local increase in sea level relative to the land at a specific point on the coast, measured by a sea-level or tide gauge. The gauge measures the combined effect of absolute SLR for the regional ocean and local/regional vertical land movement (subsidence or uplift) of the landmass the gauge sits on. Consequently, local SLR varies between coastal regions within Aotearoa New Zealand — not surprising, given that our landmass is subject to active tectonic processes and subsiding sedimentary or peat basins (e.g. the Hauraki Plains). Subsidence further compounds the local rise in sea level. As a simple example, a persistent land subsidence of 2 mm per year (mm/yr) in tandem with a rising ocean level of 4 mm/yr means the local or relative sea-level rise would be 6 mm/yr.

A common misunderstanding is that land-based sea-level gauges are not suitable for measuring the 'actual' SLR at a location or region because the record is affected by vertical land movement and localised ocean processes. However, it is the local or relative SLR, directly measured by the tide gauge, that needs to be adapted to locally (not the absolute rise of the ocean). For example, if the land (including buildings and assets on that land) is subsiding, the local SLR will be higher than the absolute rise or vice versa for land that is being uplifted by tectonic processes.

Bibliography

D. Ackerley, R.G. Bell, A.B. Mullan & H. McMillan. Estimation of regional departures from global-average sea-level rise around New Zealand from AOGCM simulations. *Weather & Climate* 2013, vol. 33. pp: 2–22. https://www.metsoc.org.nz/app/uploads/2018/02/2013_331_2-22_ackerley.pdf

R.G. Bell, T.M. Hume & D.M. Hicks. Planning for climate-change effects on coastal margins. Report prepared for the Ministry for the Environment as part of the New Zealand Climate Change Programme, Publication ME 410, Ministry for the Environment, Wellington, 2001. https://www.mfe.govt.nz/publications/climate-change/planning-climate-change-effects-coastal-margins

J. Boston & J. Lawrence. Funding climate change adaptation: The case for a new policy framework. *Policy Quarterly* vol. 14, no. 2, pp 40–49. May 2018. https://ojs.victoria.ac.nz/pq/issue/view/582

L. Cagigal, A. Rueda, S. Castanedo, A. Cid, J. Perez, S.A. Stephens, G. Coco & F.J. Méndez. Historical and future storm surge around New Zealand: From the 19th century to the end of the 21st century. *International Journal of Climatology,* 2020, vol. 40, no. 3, pp 1512–1525. https://doi.org/10.1002/joc.6283

A.J.H. Clement, P.L. Whitehouse & C.R. Sloss. An examination of spatial variability in the timing and magnitude of Holocene relative sea-level changes in the New Zealand archipelago. *Quaternary Science Reviews,* 2016, vol. 131, pp. 73-101. http://dx.doi.org/10.1016/j.quascirev.2015.09.025

A.J. Dougherty & M.E. Dickson. Sea level and storm control on the evolution of a chenier plain, Firth of Thames, New Zealand. *Marine Geology,* 2012, vols. 307-310, pp. 58–72. https://www.sciencedirect.com/science/article/pii/S0025322712000783?via%3Dihub

T. Frederikse, F. Landerer, L. Caron, S. Adhikari, D. Parkes, V.W. Humphrey, S. Dangendorf, P. Hogarth, L. Zanna, L. Cheng, L & Y.-H Wu. The causes of sea-level rise since 1900. *Nature,* 2020, vol. 584, pp 393–397. https://doi.org/10.1038/s41586-020-2591-3

W.R. Gehrels, B.W. Hayward, R.M. Newnham & K.E. Southall. A 20th century acceleration of sea-level rise in New Zealand. *Geophysical Research Letters,* 2008, vol. 35, no. 2, L02717. http://dx.doi.org/10.1029/2007GL032632

A. Ginzburg. How to avoid the unmanageable and manage the unavoidable climate changes. *UN Chronicle,* 2007, vol. XLIV (2), June 2007. https://unchronicle.un.org/article/how-avoid-unmanageable-and-manage-unavoidable-climate-changes

M. Haasnoot, S. van't Klooster & J. van Alphen. Designing a monitoring system to detect signals to adapt to uncertain climate change. *Global Environmental Change*, 2018, vol. 52, pp 273–285. https://www.sciencedirect.com/science/article/pii/S095937801830445X

D. Hikuroa. Mātauranga Māori and its role in coastal management. In S. Morgan, D. Neale & C. Hendtlass (Eds), *Coastal Systems & Sea Level Rise*. A Special Coastal News Publication 4, NZ Coastal Society, December 2020.

IPCC (2019b). Technical Summary. In: *IPCC Special Report on the Ocean and Cryosphere in a Changing Climate*. [H.-O Pörtner, D.C Roberts, V. Masson-Delmotte, P. Zhai, E. Poloczanska, K. Mintenbeck, M. Tignor, A. Alegría, M. Nicolai, A. Okem, J. Petzold, B. Rama, & N.M. Weyer (Eds.)]. https://www.ipcc.ch/srocc/chapter/technical-summary/

IPCC (2019a). Summary for Policymakers. In: *IPCC Special Report on the Ocean and Cryosphere in a Changing Climate*. [H.-O Pörtner, D.C Roberts, V. Masson-Delmotte, P. Zhai, E. Poloczanska, K. Mintenbeck, M. Tignor, A. Alegría, M. Nicolai, A. Okem, J. Petzold, B. Rama, & N.M. Weyer (Eds.)]. https://www.ipcc.ch/srocc/chapter/summary-for-policymakers/

D.J. King, R.M. Newnham, W.R. Gehrels & K.J. Clark. Late Holocene sea-level changes and vertical land movements in New Zealand, *New Zealand Journal of Geology and Geophysics*, 2020. https://doi.org/10.1080/00288306.2020.1761839

R.E. Kopp, R.M. Horton, C.M. Little, J.X. Mitrovica, M. Oppenheimer, D.J. Rasmussen, B.H. Strauss & C. Tebaldi, Probabilistic 21st and 22nd century sea-level projections at a global network of tide-gauge sites. *Earth's Future*, 2014, vol. 2, no. 8, pp. 383–406. http://dx.doi.org/10.1002/2014EF000239

S.A. Kulp & B.H. Strauss. New elevation data triple estimates of global vulnerability to sea-level rise and coastal flooding. *Nature Communications*, 2019, vol. 10, Article 4844. https://doi.org/10.1038/s41467-019-12808-z

J. Lawrence, R.G. Bell, P. Blackett, S. Stephens & S. Allan. National guidance for adapting to coastal hazards and sea-level rise: Anticipating change, when and how to change pathway. *Environmental Science and Policy*, 2018, vol. 82, pp. 100–107 https://doi.org/10.1016/j.envsci.2018.01.012

J. Lawrence, J. Boston, R.G. Bell, S. Olufson, R. Kool, M. Hardcastle & A. Stroombergen. Implementing pre-emptive managed retreat: Constraints and novel insights. *Current Climate Change Reports 2198-6061, Progress in the Solution Space of Climate Adaptation* (E. Gilmore, Section Editor), Springer, 2020. https://link.springer.com/article/10.1007/s40641-020-00161-z

LGNZ. Community engagement on climate change adaptation: Case studies. Report by Local Government NZ, August 2020. https://www.lgnz.co.nz/our-work/publications/case-studies-community-engagement-on-climate-change-adaptation/

V.A.W.J. Marchau, W.E. Walker P.J.T.M. Bloemen & S.W. Popper (Eds.) *Decision Making under Deep Uncertainty: From theory to practice.* Springer International Publishing, 2019. https://www.springer.com/gp/book/9783030052515

MfE. Coastal Hazards and Climate Change: Guidance for Local Government. R.G. Bell, J. Lawrence, S. Allan, P. Blackett, & S.A. Stephens, (Eds.), Ministry for the Environment Publication No. ME 1292. Wellington: MfE, 2017. http://www.mfe.govt.nz/publications/climate-change/coastal-hazards-and-climate-change-guidance-local-government

MfE. National climate change risk assessment for New Zealand (Arotakenga Tūraru mō te Huringa Āhuarangi o Āotearoa) – Snapshot. Ministry for the Environment Publication No. INFO 955. Wellington, MfE, 2020. https://www.mfe.govt.nz/climate-change/assessing-climate-change-risk

MfE and Hawke's Bay Regional Council. Case Study: Challenges with implementing the Clifton to Tangoio Coastal Hazards Strategy 2120. Ministry for the Environment – Hawke's Bay Regional Council partnership project. Ministry for the Environment Publication No. INFO 934. Wellington, MfE, April 2020. https://www.mfe.govt.nz/publications/climate-change/challenges-implementing-clifton-tangoio-coastal-hazards-strategy-2120

MfE/StatsNZ. Coastal sea-level rise. Wellington, MfE/StatsNZ, 2019. https://www.stats.govt.nz/indicators/coastal-sea-level-rise

Ministry to the Environment, see MfE.

M. Oppenheimer, B.C. Glavovic, J. Hinkel, R. van de Wal, A.K. Magnan, A. Abd-Elgawad, R. Cai, M. Cifuentes-Jara, R.M. DeConto, T. Ghosh, J. Hay, F. Isla, B. Marzeion, B. Meyssignac & Z Sebesvari. Sea level rise and implications for low-lying islands, coasts and communities. In *IPCC Special Report on the Ocean and Cryosphere in a Changing Climate* [D.C. Pörtner, Roberts, V. Masson-Delmotte, P. Zhai, E. Poloczanska, K. Mintenbeck, M. Tignor, A. Alegría, M. Nicolai, A. Okem, J. Petzold, B. Rama, N.M. Weyer (Eds.)] IPCC, 2019.

R. Paulik, S.A. Stephens, S. Wadhwa, R.G. Bell & B. Popovich. National-scale built-environment exposure to 100-year extreme sea levels and sea-level rise. *Sustainability*, 2020, vol. 12, no. 4, p. 1513 (Special issue 'Understanding and Preparing for Uncertainty in Sustainable Disaster Risk Management'). https://doi.org/10.3390/su12041513

R. Paulik, S.A. Stephens, S. Wadhwa, R.G. Bell, B. Popovich & B. Robinson. Coastal flooding exposure under future sea-level rise for New Zealand. NIWA Client Report 2019119WN prepared as part of the National Flood

Risks & Climate Change Project for the Deep South National Science Challenge, 2019. https://www.deepsouthchallenge.co.nz/projects/national-flood-risks-climate-change

PCE. Preparing New Zealand for rising seas: certainty and uncertainty. Wellington: Parliamentary Commissioner for the Environment, 2015. Retrieved from www.pce.parliament.nz/publications/preparing-new-zealand-for-rising-seas-certainty-and-uncertainty

R. Peart. *Castles in the Sand: What's happening to our New Zealand coast?* Craig Potton Publishing, 2009. https://www.eds.org.nz/our-work/publications/books/castles-in-the-sand-whats-happening-to-the/

H.L. Rouse, R.G. Bell, C. Lundquist, P. Blackett, M.D. Hicks, D.-N. King. Coastal adaptation to climate change in Aotearoa-New Zealand. Review Article. *NZ Journal of Marine & Freshwater Research*, 2017, vol. 51, no. 2, pp. 183–222. doi: 10.1080/00288330.2016.1185736. First published online: 13 July 2016

S.A. Stephens, R.G. Bell & J. Lawrence. Developing signals to trigger adaptation to sea-level rise. *Environmental Research Letters*, 2018, vol. 13, no. 10, article 104004. https://iopscience.iop.org/article/10.1088/1748-9326/aadf96

J. Stephenson, J. Barth, S. Bond, G. Diprose, C. Orchiston, K. Simon & A. Thomas. Engaging with communities for climate change adaptation: Introducing community development for adaptation. *Policy Quarterly*, May 2020, vol. 16, no. 2, pp. 35–40. https://ojs.victoria.ac.nz/pq/article/view/6480

M.J. Widlansky, X. Long & F. Schloesser. Increase in sea level variability with ocean warming associated with the nonlinear thermal expansion of seawater. *Communications Earth & Environment*, 2020, vol. 1, no. 9, pp. 1–12. https://doi.org/10.1038/s43247-020-0008-8

P.L. Woodworth, M. Menéndez & W.R. Gehrels. Evidence for century-timescale acceleration in mean sea levels and for recent changes in extreme sea levels. *Surveys in Geophysics*, 2011, vol. 32, pp. 603–618. https://link.springer.com/article/10.1007/s10712-011-9112-8

The impact of climate change in Aotearoa New Zealand on native flora, fauna and biodiversity

MATT MCGLONE
Environmental and climate change researcher

Global climates have always been changing. Over the past 2 to 3 million years in New Zealand, short, warm interglacial periods, during which temperatures have been as much as 2–3°C warmer than now, have alternated with long glacial periods, with cooler temperatures that have been as much as 5°C lower than today. Over all time scales, from days through to millennia, temperatures have fluctuated. The plants and animals which inhabit New Zealand are therefore supremely well adjusted to these changes and, all things being equal, should be able to accommodate the predicted warming of the next few decades.

But things are no longer the same. Aotearoa New Zealand was first settled in the thirteenth century by the voyagers from eastern Polynesia who were to become Māori, and then in the eighteenth century, by waves of settlers from all over the globe. Human pressure has shifted the balance decisively against the native biota.

Māori brought with them kiore (Pacific rats), which ate their way through a good portion of the vulnerable bird, reptile and insect fauna. Within a couple of hundred years people had hunted larger birds to extinction and burned more than half the lowland forests. A new and almost entirely indigenous equilibrium was established, consisting of a partly forested landscape stripped of many of its larger animals. But it was not to last long. With the arrival of Europeans, New Zealand, after millions of years of isolation, has been rapidly incorporated into a globalised world. New Zealanders have enthusiastically adopted the economic and cultural norms of this wider global community and, as a consequence, impoverished its indigenous biota.

Our birdlife and fresh-water fish had few defences against novel predators, and many plants are vulnerable to browsing by deer, thar, goats, possums and rats.

The drivers of this hollowing-out and degradation of native biodiversity are clear. Introduction of mammals to an archipelago free of native terrestrial mammals, and of new predatory fish to its waterways, has fundamentally changed the ecology of nearly all indigenous ecosystems. Our birdlife and freshwater fish had few defences against novel predators, and many plants are vulnerable to browsing by deer, thar, goats, possums and rats. The onslaught by vertebrate pests grinds on: the most vulnerable native biota vanished quickly, while others are being slowly worn down, their populations becoming smaller, more fragmented and exposed. Browsing and disruption of pollinator and seed-disperser networks have stalled regeneration of some trees, favouring others and thus altering the original make-up of the forests and shrublands. The arrival of exotic weeds — accidental or as escapees from cultivation — has completely altered the composition of many wild landscapes and river beds. Pines, broom, gorse, hakea, brome; the list can be extended almost indefinitely. They all take advantage of the opportunities offered by clearance and fire to outcompete or even eliminate their indigenous rivals.

Insect pests, not prominent in the first wave of animal introductions, have accelerated their colonisation over the past 50 years or so, with the advent of fast shipping and mass air-transport. Just to give one example, aside from the early introduced and relatively benign honey bee, New Zealand had few social insects in the mid-1800s. Now wasps are ubiquitous, reaching plague proportions in our forests, and ants are spreading rapidly. Finally, fungal and viral pathogens — some accidentally introduced, some self-introducing — are threatening forest, coastal and dryland communities; kauri, the native myrtles such as pōhutukawa, and coastal and inland cress (*Lepidium*) species being just the most recent victims.

As well as this new and damaging biotic influence, we have the direct effects of human activity. Clearance of native forests largely stopped around the turn of this century, but not so the elimination of native non-forest communities through intensification of agriculture. The farm landscapes of the first half of the twentieth century, created by fire, horse-drawn machinery and hand labour, while clearly destructive of native ecosystems, left wild, uncultivated spaces along roadsides and hedges, and in wetlands and patches of scrub and bush. These old farm countrysides had a charm of their own, with flowering hedgerows, shelter belts of macrocarpa and pine, patches of native scrub, tussocks and bush, and raupō- and flax-ringed ponds.

Since the 1940s, mechanisation and agricultural chemicals have systematically reduced not only surviving or remnant native ecosystems but also the diverse exotic wild communities that often sheltered many native species. In their place extend extraordinary agricultural biodiversity deserts, carpeted with nitrogen-boosted grass. Residues of the chemicals necessary to keep functional the plants and animals that fuel these artificial factory farms often persist in the soils, in the mud of our lakes, and even in our honey. The powerful insecticides and hormone sprays in particular may be having a pervasive effect down the food chains that sustain a myriad of living things.

Along with land-use intensification has come increased demand for water abstraction and power, resulting in unnatural rivers, dammed in their higher reaches and deprived of natural flow and hemmed in by stop banks in their lower reaches. Spread of urban areas has permanently converted large areas of soil to housing, and development along shorelines has hardened the coastline, eliminating sand-dune fields and constricting and filling in estuaries.

While the drivers of this massive impoverishment of the natural world vary in intensity from country to country, the trajectory is the same everywhere. Unchecked human population and economic growth each year increases the proportion of natural resources — land, water and energy — sequestered for human uses. New Zealand is not overpopulated by global standards but we produce enough food to feed 40 million. More for us, our farm animals, our crops; less for every other living thing. We are in the midst of the sixth great extinction event that the Earth has experienced, with unprecedented and accelerating loss to the natural world.

It is against this bleak background that the impact of climate change in New Zealand needs to be seen.

The impact so far

Climate affects plants and animals in two fundamental ways. First, extremes of temperature or water availability may kill organisms outright. Second, the average climatic conditions over a year and their seasonal distribution will greatly affect growth and reproduction and, depending on competitors, change the abundance of species. These two fundamental factors combined determine the abundance and distribution of species, all other non-climatic factors being equal. But they rarely are.

Soils play a very important secondary role, especially in controlling the distribution of specialist organisms that require or tolerate certain types of substrates such as limestone, peat or rock outcrops. Disturbances such as slips, blow-down during storms, flooding or fire provide niches for plants and animals that may otherwise be excluded by long-lived vegetation cover.

History also plays a large, but not well understood role. Plant or animal species may have been excluded from an area by ice or cold temperatures during the last glaciation, for example, or by the extensive volcanic-ash showers that have blanketed the central Volcanic Plateau, or even events further back in time. When all these factors exert pressure on landscapes transformed by human actions, it is difficult to know exactly what is driving changes in species behaviour, abundance or distribution.

To this uncertainty around the reaction of plants and animals to climate change, we have to add in the effect of New Zealand's highly oceanic climate. Being narrow and surrounded by oceans means that the New Zealand landmass is subjected to ocean weather with little of the amplification seen in larger landmasses, where the land surface heats up greatly relative to the adjacent oceans in summer, but cools markedly in winter. Instead, New Zealand seasonal cycles tend to be moderate, and day-to-day fluctuations have an outsized influence. This seasonal uncertainty in weather may have led to the native plants and animals having an inbuilt resilience in the face of such variation. Our native insects, for instance, by and large lack a profound winter diapause or hibernation period, as the winters here are never cool enough to demand it. Our reptiles likewise remain active under cool daily temperatures that elsewhere in the world would leave them in torpor. Few native plants are annuals — a strategy to avoid harsh winters or drought. Only in the central southeast South Island do we find them in any abundance, a consequence of this being the most continental of all the New Zealand regions. Deciduous native plants are also uncommon. The best strategy for nearly all native plants is to have moderately hardy over-wintering leaves.

New Zealand's average annual temperatures have risen by more than 1°C in the past 100 years. Most species and communities are directly or indirectly controlled by climatic factors, and so in response to this, we might expect many to have shifted their vertical range by about 180 m and latitudinally perhaps by about 200 km, to adapt to this temperature change. But unlike many other areas of the world, there has been little documented alteration in the distribution or abundance of New Zealand native plants or animals that can be unambiguously attributed to climate change.

New Zealand's average annual temperatures have risen by more than 1°C in the last 100 years.

We first have to acknowledge that we could be missing a lot because we have not been looking too hard. There are few long data series. We know little about insects, for instance. But even with well-understood groups such as birds, we have little systematically collected data aside from some special sites or species under management. Even regarding birds, detailed data-sets extending back to the last major climate warming in the 1950s are rare. In the case of plants, there are well-documented cases of species previously confined to northern districts — such as karaka and pōhutukawa — colonising areas much further south, but in all cases these trees were transported south by humans before spreading into the wild. They may indicate a response to climate change, in particular reduced frost frequency in the south, but we cannot be at all sure.

This is not to say that there have been no climate-related events. For instance, mass die-offs of pīwakawaka occur from time to time during cold winters. Drought has caused forest dieback in some places, often when dry conditions have struck normally moist regions such as the upper slopes of mountains or the windward lowlands. A spectacular two-week-long cold snap with record below-freezing temperatures in Southland and Otago at the beginning of July 1996 killed or damaged many mature native trees, some hundreds of years old.

Drought has caused forest dieback in some places, often when dry conditions have struck normally moist regions such as the upper slopes of mountains or the windward lowlands.

Some of these events might seem counter-intuitive, as they involve biodiversity loss through cold events during a prolonged warming trend. However, the biggest conundrum concerns tree line: the altitude above which forest cover cannot grow. Globally, there is a great deal of evidence for movement upward of tree lines from the mid-twentieth century onwards. While most tree species in New Zealand would have to migrate latitudinally at the rather rapid (and, for most, impossible) pace of 2 km per year to keep up with the 1°C change since 1909, tree lines would have to rise only about 150 m vertically, thus moving horizontally perhaps less than 5 m per year. This is well within the normal dispersal range of most trees.

It is therefore surprising that New Zealand tree lines have not shifted at all in over 200 years. Beech are the most predominant species at the tree line, creating a spectacularly level boundary between forest and low shrubs or tussock grassland. Photographs from the late nineteenth and early twentieth century show that these well-defined tree lines were more or less exactly where they are now.

This is certainly unusual, as annual temperatures were definitely much cooler in the early twentieth century than now. However, detailed studies of beech tree lines have shown that their annual diameter growth has been largely unresponsive to summer temperature changes. It could be argued that lack of seed at the tree line is responsible, but the forests just below tree line seem to have no problem regenerating, which rules out that explanation.

In looking for explanations for the static tree line, we first have to take into account that biological processes are very slow at altitude due to cold, low nutrition and slow soil formation. Lags of a decade or longer are likely. Furthermore, trees are long-lived and, being immobile, once they have germinated they have to survive for many years in the same spot. A warm summer may allow cold-sensitive seedlings to establish, but if harsh winters and cool summers follow, they may not thrive. So, if warming climates are to induce permanent changes, the warming has to be major so that characteristic day-to-day weather variability no long creates a risk.

The period from 1860 to 1950 in New Zealand was about 1°C cooler on average than the late twentieth century average, but much colder in winter (which does not affect tree line) than during the growing season. Given that the majority of trees at the tree line are around 150–300 years old, it is entirely possible that the cohort established before 1860 did not die back with the onset of cooler climates, as well-established adults can endure more stress than seedlings.

Moreover, temperatures in the southern South Island have not warmed at anything like the rate of those in the Far North. Dunedin has seen a century-long trend of 0.62°C warming, while Auckland has warmed over the same period by 1.2°C. The tree lines of the Southern Alps have therefore probably experienced less warming than the New Zealand-wide norm.

And, finally, year-to-year variations have dwarfed the trend line: within any given decade, mean annual temperatures can vary by up to 2°C. Since the 1950s, southerly airflow has strengthened over New Zealand, to some extent counteracting the influence of warmer sea surfaces in the southwest Pacific. Our conclusion is that no year has been cold enough since the 1850s to cause a tree-line retreat, but there has also not been a sufficiently long run of warmer-than-normal years to permit the sensitive seedlings of long-lived trees to establish in exposed sites at higher altitudes.

Some alpine insects may have had a much faster reaction over the same period. The end-of-season snow line has been retreating rapidly over the past 40 or so years. Insects dependent on the specialised snow-line habitat (snow- and ice-covered most of the year, with only a short period clear of snow, at the height of summer) are naturally well adapted to track this narrow zone, and have done so. A warming climate and shorter snow-lie should also have created conditions suitable for thickening and upwards extension of alpine shrublands and grasslands and their animal biota as well; however, there is no evidence that they have done so, although heavy mammal-grazing pressure makes it difficult to be sure.

Another example is the mangrove. Mangrove needs warm water and high minimum temperatures to thrive, and it is abundant along the coastline of the North Island south to Kawhia in the west and the Bay of Plenty in the east. It has been spreading within its range, most likely because of the influx of muddy sediments into its estuarine habitat, and has therefore become a major issue in northern districts for boat owners and those who have paid good money to buy a seascape. Warmer, less frosty climates have probably helped its vigour. But despite warming sea temperatures and much higher minimum temperatures, it has not spread south. Here, the lack of suitable habitat in the form of muddy estuaries immediately to the south of its range might be the explanation for the lack of movement of an otherwise highly mobile species with water-transported fruit.

If the next 1°C rise in temperature over New Zealand was to bring with it a similarly low level of change as the last 1°C rise, we could perhaps relax — at least as regards biodiversity. But that is highly unlikely. A 1°C rise will lift the trend line above the year-to-year variation that seems to have been dampening substantial change; 2°C further and we will be in uncharted territory. Barring some technological fix, we are going to have to deal with a much warmer New Zealand, and sooner or later very substantial biodiversity change.

What change may bring

The New Zealand biota has over 30,000 named terrestrial species, each with its own unique ecological niche, but the true number is thought to be twice this. As we have seen, even a highly significant shift in temperature may not be easily detected against the background of all the other ongoing changes that affect our plants and animals. Even when a species is highly responsive to a shift in climate, other factors may prevent its movement or accommodation.

Perhaps the best we can conclude from our very recent history is that the situation is complicated. We can use models to help us look into the future, but they are mostly based simply on observations of present-day distributions and have no dynamic or interactive component. Unlike the general circulation models used for climate change forecasting, they are really just a sophisticated sort of guesswork when used in prediction mode. Some work has been carried out in which experimental plots are warmed to see the consequences for plant and animal life, but they are small-scale and subject to many constraints.

Fossil evidence from periods in the past that were warmer than now can give us some clues. The most recent of these was around 10,000 years ago, when New Zealand's annual temperatures were possibly as much as 1.5 to 2°C higher than in the mid-twentieth century. Oceans were much warmer around the south of New Zealand. As a result, southerly to westerly winds were weaker, and cloudy, humid, often frost-free conditions prevailed in the west. Weaker circulation meant fewer rain-bearing storms and drier eastern districts. Central Otago and inland Hawke's Bay were too dry for forest. Frost-sensitive plants were widespread, mangrove was found further south and cold-tolerant beech was rare. Surprisingly, tree lines were no higher than now, although these alpine forests were very different in composition, as they consisted mainly of broadleaved scrub and small conifers.

Prediction of how the whole species ensemble may react to climate change over the next few decades, therefore, relies on a mix of past behaviour, ecological models, experimentation and guesswork, based on a few general principles. We need to use all of these methods to future-cast. These will be far better than predictions of what might happen to the human community in the same time span, but still need to be treated with some balanced scepticism. And we must remember, the most unlikely outcome is no substantial biological change. A 2°C rise in temperature by the end of this century, with its associated shifts in wind, rainfall, drought, frost and cloudiness, is the equivalent of one-quarter of the typical shift from cold glacial to warm interglacial period. On top of this we must factor in the extraordinary rise in atmospheric carbon dioxide to levels not seen for 50 million years. In what follows I will look at the likely impact on major ecosystems.

First, a biogeographic rule

The larger an area covered by an ecosystem, the more species it contains. To take an example, a small plot of 500 m^2 within a northern lowland forest may have up to 20 tall shrub and tree species but there are 10 times that number in those forests as a whole. As we increase the number of plots, the number of species encountered grows, but at an ever slower rate. Roughly speaking, if we sample 10 per cent of the area, we will be pretty certain of finding 30 per cent of the total species. When 50 per cent of the area has been covered, all but a handful of rare species will be included.

This tendency for numbers of species to initially increase rapidly with area and then to slow but keep growing ever larger, applies across scales. The resulting relationship between species number and area can then be used to estimate what may happen if the area covered in a given ecosystem is whittled away. This is exactly what has happened with lowland ecosystems in New Zealand. They are now greatly reduced in extent, and fragmented.

Surprisingly, little species loss can so far be directly attributed to this drastic clearance. While in part this may be due to the resilience of the native plants and animals, most believe that it is but a matter of time before a slow erosion of species from bush patches leads to a marked national loss of species. That is, we already have an 'extinction debt' — the walking dead. Climate change is likely to advance this day of reckoning.

> **Of all the changes that we are going to face in the next few decades, rising atmospheric CO_2 levels are the most certain.**

Much higher atmospheric carbon dioxide level

Of all the changes that we are going to face in the next few decades, rising atmospheric CO_2 levels are the most certain. Long geological records show that biodiversity has waxed and waned in response to CO_2 fluctuations, and that most of this change is due to climatic shifts induced by alterations in CO_2 levels.

Nevertheless, CO_2 is the feedstock for all plant life, and experimentally it has been shown that increases in its concentration markedly boost plant growth. With higher concentrations of atmospheric CO_2, most plant species photosynthesise more, increase their growth and use less water, and their leaves contain lower concentrations of nitrogen and protein. Raised CO_2 levels have other positive impacts on plants, particularly trees. More CO_2 increases the optimum temperature at which they operate, making them capable of increased growth under dry conditions and helping to proof them against heatwaves. They are not rendered any more susceptible to insect pests, and perhaps less so, because they are producing less nutritious foliage.

These positive effects are lessened under cool conditions on nutrient-poor soils. Thus the major protective effects will be seen in lowland and drought-stressed situations. That said, individual plant species will have markedly variable responses, and it is inevitable that some will do better than others, which will lead to compositional changes. On the other hand, positive responses to increases in atmospheric CO_2 will greatly increase the ability of plants to persist where they are despite higher temperatures.

Disruption of biological networks

Many plants and animals are cued into reproduction cycles by regular seasonal changes. Sometimes these are controlled by day length, as in many plants; sometimes by temperature or rising water levels or some other more subtle alteration of the environment. Given that most ecosystems have hundreds of species interwoven into a network of interactions, the potential exists for these seasonal cues to become disrupted.

For example, a plant may flower before its preferred pollinator is mature, and therefore not bear as heavy a fruit crop. Increased temperatures may change sex ratios, as happens with tuatara, where warmer soils lead to more male eggs developing. Migratory birds may arrive after biological productivity has already peaked in their chosen habitat.

An important point to note here is that most of these effects are hypothetical. The inherent variability of the New Zealand climate pretty much guarantees a built-in ability of native plants and animals to accommodate change. In those cases where climatic variation exceeds this inherent capacity, the change itself will exert a powerful genetic effect on the population in the direction of adaptation. That these species have survived similar shifts in climate in the past suggests that at least until fundamental climatic limits are exceeded, ecosystem disruption by itself will not be a major factor.

Coastal systems

Along with unprecedented increases in carbon dioxide, rising sea levels are a given. By the end of the century, the average sea level around our coasts will have risen by somewhere between 0.3 and 1.0 m, with the likeliest figure being around 0.5 m. This is very bad news for natural coastal ecosystems, because they are already among the most weed-infested, disturbed and fragmented in the country. Not only that, but also the human community now prizes dwellings along the coastline and demand for marine recreation facilities has surged. In the resulting social struggle for a foothold along the coastline, the needs of natural ecosystems have been neglected.

The problem is most acute on low-lying, soft muddy or sandy coasts, and is most critical along estuaries. Our major cities are built close to estuaries, which have been subject to reclamation and a hardening of their edges by roads, installations and dwellings. As sea level rises, the estuarine vegetation zone is compressed up against this barrier. Continuing accelerated sedimentation from development along waterways will continue to fill in many estuaries, and, as is seen throughout the north, this will lead to undesirable trends such as the unchecked expansion of mangrove or some vigorous weeds such as *Spartina* (cord grass). The scenario for sandy exposed coasts is not much better. Higher sea levels along soft rock- and dune-edged coastlines greatly exacerbate the impact of storm events.

Major storms — and the frequency of these has been predicted to increase markedly — are accompanied by low pressure, which of itself raises sea level, and giant waves then operate off the higher base level. The results throughout New Zealand are already clear: many coastal communities have suffered loss of land, roads and dwellings. In part this is because of heedless development into high-risk coastal areas over the past 30 years, despite regular warnings. However, houses can be relocated; not so the dune systems, which are already subject to destructive erosion by vehicles, unchecked growth of weeds such as marram and lupin, and dune bulldozing to improve sea views; the development of farmland and housing estates behind the narrow dune fringes leaves no room for expansion. The most likely scenario here has already played out around New Zealand. Rapid-growing, sea-spray-resistant pines are used to stabilise moving dunes but in turn eliminate the dune-adapted native vegetation and animals.

Lowland forests

Before human arrival, the lowlands of New Zealand were nearly entirely clad in tall forests, or tall shrubland in recently disturbed or harsh sites. About 90 per cent of this lowland forest is now gone; the only extensive areas of forest that extend from the mountains to the sea are on the west coast of the South Island. Elsewhere, small patches surrounded by agricultural land or fire-induced shrubland and grass are the rule.

Climate change per se (within the 2–3°C forecast) would have posed very little threat to these lowland forests and their animal inhabitants if humans and their exotic fellow-travellers were not present. We know this because similar warmings have happened before. While some of the canopy trees are intolerant of drought, warmer conditions largely do not seem to have any negative effects. As we have seen, elevated CO^2 levels may actually help these trees. A number of species are quite well adapted to drought, such as mataī, tōtara and kōwhai, and these trees already dominate eastern forest fragments. Predictions for the west of the country are for greater, not less rainfall; therefore the main threats are in drier eastern areas, where extreme droughts could eliminate some of the more vulnerable species or leave them open to disease and human-lit fire, which is the inevitable accompaniment of dry conditions.

The Christchurch Port Hill fires of February 2017 showed the complex threat that arises when a mixture of fire-prone exotic pines, shrubland and grassland is adjacent to dwellings, roads and native forest patches. The damage to the core forest areas was slight, although the surrounding shrubland which was harbouring tree regeneration was more badly hit. This is very likely to be the model for the near future: more-frequent fire policing the boundaries of native vegetation, preventing expansion. Similar extensive fires occur every year in the native scrub of the drier east, maintaining a continuing state we call a 'fire climax'.

Clearance of native lowland native forest has virtually ceased. Fire may nibble at the edges but likely not more; drought may lead to elimination of some vulnerable species, but not many. But the future for these forest remnants is anything but bright. Weeds are an ever-present threat, and stoats, rats, possums and wasps have had many decades to devastate both fauna and flora. Warming may expose the forest to new weed pests. There are hundreds of potential warm-climate weeds in the wings, some of which appear to occupy ecological niches poorly served by native trees. For instance, the nitrogen-fixing small tree *Morella faya* of the Canary Islands has devastated large forest tracts in the Hawaiian islands, its bird-dispersed fruit enabling it to get a foothold on less-fertile soils and then transform them. It is present here but not yet naturally spreading.

Invasive palms are another subtropical threat already moving from gardens and parks into nearby forests in the north, as it shifts into a permanently frost-free state. Some of these invaders may fade away after an initial spread and period of hyperabundance, as natural enemies catch up with them, but we would be unwise to count on this.

Lowland wetlands

Two major types of wetlands occur in the lowlands: extensive wet swamps and fens, where the water sits above the surface of the peat for at least part of the year; and domed bogs, where the water table is perched below the surface. Only a fraction of the original lowland wetlands remain in New Zealand, many having been drained for the valuable agricultural soils they occupy, and only a very small fraction of these remaining swamps are in a healthy state.

Swamps are highly sensitive to environmental changes, as they act as lowland sumps. Waterways bring in pollutants and excess nutrients; drainage ditches lower the water table; fire spreads easily across

Swamps are highly sensitive to environmental changes, as they act as lowland sumps.

them as there are no barriers to prevent spread among tall summer-dry reeds, sedges and shrubs; weeds invade easily; and stock trample their margins. Climate change will exacerbate all of these effects, increasing fire and drying surfaces during severe drought years.

On the other hand, many high-domed bogs have been largely untouched by pollutants as they receive nearly all their water via rain and are quite resistant to weed invasion because of the very low nutrient levels in the peat. However, continual encroachment by drainage of adjacent land has shrunk their extent, drying their edges and making them vulnerable to weeds.

Drought once again is the major climate change threat here. The vegetation on the surface survives solely on rainfall, and drawdown of the water table results in a waxy, water-repellent surface which makes seedling survival difficult. Fire then has the potential to create permanent vegetation-free patches, with loss of peat through oxidation. Even with the best management possible, domed bogs may be in for a rather dismal time.

Shrublands, fernlands and grasslands

Originally the New Zealand lowlands were covered in forest, but now huge extents are in mixtures of low secondary forests, fern and native tussock grasses. While not the original vegetation, these native communities are still valuable and harbour a richness of plant and animal life. The tussock grasslands in particular are part of a much-loved iconic New Zealand landscape.

Centuries of fire have created virtual fire climaxes, in which each fire reinforces the dominance of fire-resistant or fire-tolerant native species. Some of these — for instance, bracken and mānuka — resist fire due to their buried stems or roots or thick barks. Mānuka is one of the few plants in New Zealand that have a particular strategy for recovery after fire: its large, woody seed capsules are often held unopened on the branches for more than a year after flowering, but open to release seeds when subjected to the heat of a fire. Kānuka has small capsules and releases seed in the autumn, but the seeds are very small and wind-dispersed, and colonise burnt ground very well. Frequent fires thus maintain the dominance of these native species.

The real problem is that many exotics such as hakea and gorse are also superbly well adapted to fire, and the frequent fires they encourage eliminate most fire-sensitive natives. The game-changers, however, are the fire-adapted exotic pines: in particular *Pinus contorta* and *Pinus nigra*, which are now spreading exponentially into shrubland and grassland in eastern districts, fire or no fire. Fast-growing even on poor soils and under harsh climates, these species have the potential to completely alter montane and lowland largely native communities into near monocultures. Nearly as significant are the rhizomatous, turf-forming exotic grasses which choke out native herbs and prevent regeneration of native shrubs and trees. Against this background of ongoing exotic invasion and drought-induced fire, other effects of climate are likely to be minimal in contrast.

Montane and alpine forest

Much of our remaining native vegetation cover is in the mountainous, wet hinterland. Protected by climate from Māori and settler fire, and then subsequently by steep terrain and poor soils from conversion to agriculture or exotic forestry, they are the regions least affected by exotic weed invasion. Nonetheless, mammalian pests have penetrated throughout, right up into the alpine zone, wreaking havoc on birds, reptiles, large insects and land snails. In the beech-covered regions where wasps have reached plague proportions, the effect on larger invertebrates — spiders, moths and flies in particular — has been devastating.

The cloudy, wet prevailing climate of this zone is likely to moderate the effects of climate change, in particular in the west, where rainfall is predicted to increase. However, the strong trend towards warmer, less-frosty winters that we have already experienced is likely to work in favour of mammalian pests and wasps, through increasing survivorship over the winter, while somewhat warmer summers may encourage more and more regular fruiting, which will favour rats and mice. More rats provide more food for stoats, which are savage predators of birds and reptiles.

A widespread phenomenon in New Zealand is heavy fruiting of certain types of trees (mast seeding), followed by several years of low fruit production. The beeches and rimu are the best-known mast seeders, but there are many trees which follow this pattern. Mast fruiting has long been known to be triggered by a warmer-than-normal year before the fruiting episode, and it was expected that warming would result in more-frequent and heavier seed-fall and thus more pest-plague years in the forests. However, more-recent studies show that it is the *difference* between the warmth of years, not the absolute warmth of a given year, that drives the cycle. Increased variability in the warmth of the summer season and not increased warmth per se may be the important factor. However, increased temperature variability will probably have a relatively subtle effect, most likely by inducing less-frequent but heavier fruiting years.

The alpine zone

The alpine zone is relatively young, having come into existence as a permanent feature but a few million years ago. Rapid evolution of species within the zone has created some of the most spectacular plants and animals of the entire biota. Some are unique: nowhere else in the world are there alpine parrots and geckos. It is therefore unfortunate that the most dire, and perhaps most credible, predictions of ecosystem change and species loss concern the alpine zone. There are a number of good reasons why the effects of climate change should be more fundamental and injurious to biodiversity in the alpine zone.

The alpine zone is uniquely determined by temperature. Within the relatively narrow band of 500–600 m of vertical altitude between the tree line and the end-of-summer snow line, distinctive bands of vegetation are packed, from tall shrubland through tussock grassland to fell field of cushion plants and herbs, and finally scattered herbs, prostrate woody plants and bare gravel and rock. Each zone has its own suite of cold-adapted animals, including geckos, skinks, grasshoppers, wētā, beetles, butterflies, moths, bees, flies, spiders and numerous other invertebrates. Some escape the cold winter by sheltering under snow or below the soil surface; others have anti-freeze properties, and can resist freezing or tolerate some degree of freezing overnight and then recover during the day. New Zealand skinks have evolved live births as a way of avoiding cold stress. Some of the few herbs that have leaves only in summer occur here, such as the iconic Mount Cook buttercup.

The alpine habitat is complex. Poor drainage coupled with low temperatures leads to large areas covered with saturated soils, and bogs form. Exposed bluffs and rock faces abound and are subject to extreme temperature fluctuations. Due to New Zealand's relatively low latitude setting and variable oceanic weather, snow comes and goes, and many sites have large soil-surface temperature variations all year round.

It is also an extremely dynamic region, not the least because of continuing rapid uplift with associated earthquakes. In some areas, particularly in the drier eastern mountains, huge scree slides stretch from the snow line down below the tree line, and have their own specialised flora and fauna. Slips, sometimes massive like the one that removed 12 million m³ of rock and 10 m of height from Aoraki/Mount Cook in 1991, are common, as most of the alpine area is constructed from weak sandstone or jointed schist. As with alpine areas everywhere, snow avalanches scrape soil and vegetation off in long strips as they smash through forest to the valley floors. Rapid melt-back of permanent ice is likely to expose more unstable terrain to slipping.

Alpine habitat is highly vulnerable to climate change because it has two hard borders: snow and ice, and tall forest. Both of these borders will move with warming, and the summer snow line has already done so, more or less in line with the amount of temperature change. This parallel movement of the upper and lower alpine boundaries has three dramatic effects:

1. The higher a zone moves up a mountain, the less land area it occupies, because the upper levels of most mountains approximate a cone. As we have seen, all things being equal, a smaller area

translates into fewer species being present.

2. The higher up the slope, the more broken and fragmented the zone, as it moves into areas of peaks and bluffs and gravel chutes.

3. Eventually the summer snow line moves above the highest peaks, and with that a unique habitat vanishes. A 3–4°C increase in temperature will result in a vertical movement of 500–700 m — that is, a complete displacement of the zone and a loss of about half the total alpine area.

While there is no doubt that the upper boundary, the summer snow line, is retreating rapidly — by several metres a year — as mentioned above the lower boundary, the tree line, has remained static despite warming, and possibly the lower alpine vegetation zones as well. One way of viewing this is that the snow-line retreat is a strictly physical and near-linear reaction to warming, whereas the tree line is a biological response and thus decidedly more complex. As we have seen, the most likely explanation for this is the nature of the warming that has occurred, with warm years interspersed with cold years and the increasing prevalence of southerly airflow. The tree line is probably not tracking the warm years, but the cold. However, with increased warming in the next few decades, these occasional cool summers will become a thing of the past, and at that point, as elsewhere in the world, the alpine tree line should begin to move.

The upward movement will probably be slow (all biological processes are slow in the alpine zone) and the broken, steep, unstable and often saturated terrain will not permit tree growth in many places. Thus we can expect a mosaic of alpine vegetation cheek-by-jowl with stunted forest and shrubland. Many alpine open-habitat species will therefore survive. However, those that are tightly alpine-adapted with strong cold resistance, snow-sheltering habitats or brief short-summer life cycles are likely to vanish, especially as the upper snow line is extinguished.

Predictions based on a 3°C warming and a consequent rise of the zone by around 500–600 m of vertical elevation suggest an eventual loss of 200–300 plant species (about half the total alpine species) and an unknown but much larger loss of invertebrates. It seems unlikely that such a loss would occur by the end of this century, but in the longer term, it seems probable that a vast reorganisation of the alpine zone with species losses of this order will take place.

What can be done? What should be done?

When we consider this question, the issue is above all how much we value native biodiversity, unmanaged landscapes and wilderness. New Zealanders are enthusiastic about the outdoors, grassroot groups have sprung up all around the country to protect and preserve biota and ecosystems, and the news media is always ready to run with stories about endangered animals and plants. Kākāpō and kauri, for instance, have had near saturation coverage. But if we measure our commitment by how much we are prepared to forgo to allow a place for nature, we hardly value it at all, and never really have. While the underlying

At an individual level, we greatly value our native biodiversity.

factors of exotic predation, weeds and habitat loss grind on, diminishing our unique biodiversity, we put little effort into mitigation. Of the 2000 species regarded as being at risk, only 150 have management plans. Current government funding put into conservation of biodiversity is around 0.45 per cent of the total government spend. Attempts to reduce or mitigate the harm done to biodiversity by agricultural intensification have been minimal and the damage continues.

I could go on, but the point is that while we say we care, the fact is that we are collectively loath to spend on the biodiversity crisis or to rein in industries affecting it. The same applies to climate change. Despite signing up to international agreements to reduce our emissions, New Zealand greenhouse emissions continued to rise after the benchmark year of 1990, so far by nearly 20 per cent.

I mention these rather dismal statistics not to disparage the efforts that are been made — and some of these, such as the creation of predator-free islands, are world-leading. And I do not intend to suggest that New Zealanders are more heedless of their environment than other wealthy countries. I don't think we are. But what I would like to point out is that although we are in the midst of a long-standing biodiversity crisis (since the 1860s, at the very least) which shows no sign of abating, we simply do not act as though there is one. In fact, a minority are vociferous in promoting one of the key causes of our biodiversity crisis — mammalian pests — and opposing, with threats of violence, effective remedies such as 1080 aerial drops.

The lesson is that the climate crisis, which is surely with us already, will be treated in much the same way as the long-established biodiversity crisis. Unwelcome climate change effects which threaten human health and livelihood will be addressed, one way or another. Climate change consequences for biodiversity will not get any more consideration than that shown for the current extinction crisis. Given this, any recommendations for action have to be constrained.

We also have to think hard about what sort of a country we might want in the medium-term future. The world seems to be committed to at least a 2°C warming, with loss of the polar ice caps as one of the many consequences. Even if technologies are developed to suck carbon dioxide out of the atmosphere and bury it — and none seem anywhere close to being realised — there seems little possibility of the world returning to the pre-industrial norm.

However, my guess is that most New Zealanders, when looking at the climate projections for the end of the century, are not too alarmed. These would just about bring Auckland into line with the current climate of Wollongong, New South Wales. Cold-related morbidity in this country almost certainly is greater than any projected heat-related morbidity for the foreseeable future. Every winter thousands of New Zealanders seek warmer locales overseas. And there is no doubt that at a human level New Zealand is well placed to endure the climatic catastrophe that is unfolding; Bill McKibben, a leading environmentalist, has written that if the world approaches anything near the 5–6°C of warming within the realm of possibility, 'the living will truly envy the dead: this is a world where people are trying to crowd into Patagonia or perhaps the South Island of New Zealand'. We therefore accept with equanimity and perhaps even welcome a New Zealand somewhat warmer than now.

We are not alone in deciding that doing the bare minimum is the wisest course. Europe, North America, China, Australia and India have shown by their inaction that they are prepared to endure a quite substantial amount of global warming and sea-level rise if this prolongs our century-long fossil fuel spree. This means that the long climate cycles which shaped the evolution of the flora and fauna of this country, and the world, have definitively ended. There will never be another ice age to rejuvenate the landscape. The Pleistocene has ended and the Anthropocene has begun. The evolutionary trajectories of the biota have been profoundly altered and the plants and animals must adjust to a future shaped in one way or another by human interventions. Slow loss of cold-adapted flora and fauna may be part of that future reality.

We therefore accept with equanimity and perhaps even welcome a New Zealand somewhat warmer than now.

With this virtual guarantee that New Zealanders will do the minimum to prevent climate change (and in this we are in line with most of the rest of the world), let's look at what biodiversity consequences we might have to deal with. We must look at these issue by issue, as the sets of biodiversity and social issues are different in every case.

COASTAL ISSUES

The only way to deal with the impact of rising seas on coastal and estuarine habitats is to provide room for these systems to shift inland. This could not be done on a small scale. Coastal land is highly prized, especially sandy, dune-backed stretches of coast. Our population is firmly wedded to the concept of the sanctity of private property, and the personal wealth accumulated by parliamentarians, party funders and the judiciary is largely in real estate. Given that, and the propensity of those affected by compulsory acquisition of land to fight planning decisions through the courts, moving coastal areas into the public domain to give a breathing space for biodiversity is almost certain to be unpopular and unsuccessful, in the unlikely event that a political party decided to court suicide by proposing the necessary legislation. Unfortunately, the public mood is more for seawall building to prevent encroachment, and this is the very worst response of all for biodiversity. The only feasible option here is to prevent further human encroachment into coastal areas that have yet to be developed and to ban coastal defences outright.

FIRE

Fire weather and drought are both likely to increase with warming. New Zealand's oceanic climate means it has a globally low incidence of fire-inducing lightning. Fire is therefore almost entirely associated with agricultural activities or careless or deliberate firing of vegetation along roadsides. This will probably change as summer warming of the land surface will promote thunderhead formation and lightning.

Preventing fire spread is expensive, and in sparsely inhabited back-country the reaction is usually to prevent the fire reaching valuable human installations rather than limit damage to indigenous vegetation and fauna habitat. The terrain makes it difficult in any case. Freshly burnt areas are vulnerable to weed invasion, but follow-up preventative work once the fire is out is rare. Special consideration can be given to valuable forest patches threatened by fire, but more general work on fire-proofing natural areas is unlikely. The solution proffered by many in the farming community of heavier grazing to reduce fire risk simply sets the landscape on an unalterable course to less native biodiversity. We will have to get used to fire-climax communities, some of which will be dominated by pine, broom, gorse and hakea.

MIGRATION

As we have seen, a 1°C rise in temperature should result in a southward retreat of climatically determined ecological zones by 200 km or more. All things being equal, southwards movement of species is possibly one outcome. But this is where New Zealand's oceanic climate moderates outcomes. Our variable weather does not lead to exclusion through climatic extremes. For example, kauri — historically confined to the upper third of the North Island — actually grows well and fruits as far south as Invercargill. Despite the massive shifts between glacial and interglacial climates, very few cases of long-distance migration have been recorded. Instead, it appears that the most common reaction in the past has been for plant and animal ranges to shrink and fragment or expand and coalesce, according to their climatic preferences, within their current broad range. Therefore, there is every reason to believe that most of our lowland species have a similar wide range of tolerance. That is to say, the projected climate change over the next few decades is unlikely to eliminate species but rather to confine them or, alternatively, release them from constraints within their current distribution. As noted above, the major exception is where plants have been moved by humans around the country and are invading areas well outside their natural range.

A very well-studied example of this phenomenon is the native *Olearia lyallii* (a small, large-leaved coastal daisy tree) on the subantarctic Auckland Islands. Native to islands and coastal districts in the far south of the South Island, it was apparently introduced to Auckland Island by sealers before 1810. It has since spread to numerous locations around

Laurie Harbour and Enderby Island in the north of the island, invading coastal grassland and herbfields and replacing the native southern rātā forest. The question this posed for conservation managers was whether to attempt to control this invader (a costly proposition) or let it reach a natural equilibrium with the local vegetation. Studies showed that its spread had been slow, and that it appeared to be favoured only by coastal areas well within the spray zone, and in particular those areas well fertilised by sea lion dung and petrel guano. In this case, it was decided that control was not called for, because the tree fits the local ecological situation well, and seems destined to have only a limited impact on the resident vegetation.

A different proposition is the invasion of southern districts by karaka. Karaka is a medium-sized tree with large glossy leaves and fleshy fruit. Within its natural range it is not common, but to the south in mild, coastal districts it can become aggressively invasive in forest, and diminishes the abundance and diversity of other trees and shrubs. It is moderately frost-sensitive and it appears likely that the diminution of frost in recent years may have assisted its spread. The key issue is therefore not whether a southwards shift in its range is consistent with climate change (and therefore acceptable), but the impact it is having and the cost of restraining it. The equation seems tilted at the moment in favour of restraint as its large fruits are spread far and its local impact is rather large.

Issues regarding the southwards spread of undesirable New Zealand natives will probably only concern plants. While spread of insects is likely to also occur, there are no practical ways of restraining them in natural settings. Birds are simply not regarded as a problem, whether native or exotic, aside from introduced agricultural pests such as rooks or Canada geese.

A separate debate concerns relocation of plants and animals which are 'stranded' by warming climates outside of their climatic niche. The most likely areas which would fall into this category are upland habitats in the central and northern North Island. For instance, there are some apparent plant relics of previous cool climates holding on at the tops of mountain peaks in the North Island, but they seem under no immediate threat. Few of these are endemic, and it would seem to be a rather pointless exercise to relocate them before a pressing need arises or, for argument's sake, ever.

ALPINE CHANGE

Even under the most extreme scenario for the next 100 years, most of the plants and animals of lowland and montane New Zealand seem safe from climate change. We will lose some of the more vulnerable elements to pests and disease and degradation of our ecosystems through fire and weeds, but this loss was baked in from the moment Captain James Cook set foot in this country. Climate change will be but a minor contributor. However, the climatic change impact on the alpine zone is likely to be of a quite different order.

It is worth considering that over the course of the past 2 million years or so, most current alpine species have survived the impact of warmings of 2°C degrees above the present average, strongly suggesting that patches of snow and ice remained to provide sufficient habitat. But that would seem to be some sort of an ultimate limit, as it is difficult to see how an isolated, alpine endemic that can live nowhere else could persist with total loss of snow and ice, as seems certain in the mid-term future.

Translocation of species to avert their loss is often put forward as a solution and, short of reversing global warming, it is difficult to think of any other approach. New Zealand conservationists have been translocating birds, reptiles and large insects for well over a century as a response to loss of habitat to mammalian predators. Translocations within the alpine zone as isolated endemic biota run out of room is possible, but fraught with difficulties. Leaving aside the difficulties of translocating sufficient individuals to create a viable, genetically diverse population, such species would have to be moved to other similar alpine areas. But these areas will be highly likely to have closely related species already in residence, raising the chances of competitive elimination of one or other. In the future, better ecological understanding and technologies might make such translocations possible, but even so it is likely that such efforts will be limited to truly unique plants and animals, and most species will be left to fend for themselves.

Conclusion

Any predictions of the effects of climate change on biodiversity are uncertain because of our limited understanding of the current ecology of the vast majority of the roughly 70,000 species that share this country with us. We have worked out some of the rules that govern their ecological behaviour, but in biology the devil is in the details. A best guess is that for the next few decades the direct impact of climate change will be slight compared with the savage impact of agricultural intensification, damming of rivers, insecticides, fire, continuing clearance of scrub and grassland, and the largely unchecked presence of mammalian predators and herbivores throughout remaining intact native vegetation. Further threats remain in the wings such as avian malaria, new insect pests and fungal pathogens, which are difficult to guard against.

When we consider what we actually might do as a nation to avoid climatic impacts on biodiversity, looking at our past record is one of the most certain ways to predict future actions. We have rarely sacrificed economic

When we consider what we actually might do as a nation to avoid climatic impacts on biodiversity, looking at our past record is one of the most certain ways to predict future actions. We have rarely sacrificed economic opportunity for preservation of biodiversity.

opportunity for preservation of biodiversity. By and large, what we have protected is on land for which we have no other use. The most then that we are likely to do is some mitigation of the most obvious and destructive impacts.

This is not a cynical viewpoint; it is as solid a prediction as any climate prognostication. New Zealanders, as with the people of nearly every other nation, tend not to take action until unmistakable damage is done. Think about flooding, earthquakes, mine disasters, leaky homes, earthquakes, firearms control, bank regulation. Warnings are ignored, regulations weakened and then, after disaster strikes, reluctantly we take action. Then, if these actions are effective, over time our sense of danger recedes and we dismantle protections as they seem to not be needed any more.

Dramatic climate effects on biodiversity are likely to be many decades away and, for most of them, such as widespread dieback through drought in forests and shrinkage of the alpine zones, there are no obvious current solutions. Given that we are likely to live with the consequences of the twentieth and twenty-first century fossil-fuel bonanza for a few centuries to come, letting the plants and animals adapt to higher CO^2 levels, encroaching seas, warmer climates, more-variable rainfall and stormier weather actually seems the sensible thing to do.

In the meantime, as far as biodiversity is concerned, the focus should not be on climate change at all but in finding ways that the people of this country can release the pressure they are exerting on the natural world right now. This means as a first step finding out a whole lot more about what biodiversity we have, where it is and how it gets by. As regards direct action, eliminating mammalian predators will get us some of the way, but there is much more to do than that. Birds are not the only species worth saving. Insect pests may be snuffing out just as many invertebrate species or even more. We have few tools available to use against fungal pathogens which may devastate our flora. Reconciling agribusinesses with the wild landscapes so often treated as a profit problem is well overdue. Reversing the 200-year-long assault on wetlands would be a start. Coastline legislation needs to give full weight to biodiversity, so that this dynamic boundary between sea and land is no longer seen solely through the lens of property investment.

Doing these sorts of things will be the best way to build resilience into landscapes so that our native species can adapt to the new normal, and that there is actually something left worth defending when the real warming begins.

Aquatic environments under climate change in Aotearoa

ANDREW JEFFS
Professor, Institute of Marine Science,
University of Auckland

SIMON THRUSH
Professor, Institute of Marine Science,
University of Auckland

C overing just over two-thirds of the Earth's surface, our oceans have been providing a tremendous buffer to the potential effects of climate change over the past century. Around a third of the vast quantities of carbon dioxide released by humankind in the last century has been absorbed by oceans, reducing the heating effect of this gas if otherwise left in the atmosphere. The remaining greenhouse gases — those not absorbed into the oceans or taken up by terrestrial vegetation — have been accumulating in the atmosphere, resulting in an increase in the amount of the sun's heat that is trapped on Earth. To date our oceans have also been absorbing around 90 per cent of this additional trapped heat, due to the enormous capacity of our oceans to take up heat energy with a modest resulting increase in ocean temperature. Without this buffering by our oceans of both the additional carbon dioxide and the heat, our planet would already be uninhabitable. We would have no freshwater and not enough oxygen to breathe.

Despite their vast size, the Earth's oceans are changing as a result of this buffering effect. The absorption of carbon dioxide is altering the chemical nature of seawater and making it more acidic, which affects a variety of sea creatures. Absorbing additional heat is warming our oceans and melting polar ice, causing our oceans to expand in volume, leading to rising sea level along coastlines.

As the world becomes warmer, more water evaporates from the surface of the Earth, especially from our oceans. This additional moisture contributes to heavier rainfall events and more-intense storms, increasing the risk and severity of flooding and storm surge. The warming of the Earth also increases the drying of the land, so that in some places droughts have become increasingly common and severe in recent times. The increased heat energy trapped at the surface and in the atmosphere of the Earth is also altering the established weather systems, driving changes in traditional weather patterns and fuelling more-severe weather events. For example, in many places on Earth storms have become both more frequent and more intense in terms of their wind strength and the total amount of rainfall they deliver.

As an isolated chain of islands surrounded by oceans in the temperate zone of the southern hemisphere, Aotearoa New Zealand is not immune from global climate change processes, especially those occurring in our oceans. The major changes we are already seeing in New Zealand are warming ocean waters, rising sea level, changing seawater chemistry and changes in weather patterns, especially rainfall, that are affecting the terrestrial and aquatic environments. Each of these major changes has significant consequences for the environment and associated economic and cultural activities in New Zealand.

Warming ocean waters

Overall, the surface waters of the oceans around New Zealand are gradually warming, but at different rates around the country. While northeastern waters have shown little or no warming over the past 50 years, coastal seas around much of the rest of the country have been warming by 0.1–0.3°C every 10 years. These water-temperature changes appear to be driven by the shifting flow of warm ocean currents travelling across the Tasman Sea from off the coast of southeastern Australia, which is one of the fastest-warming ocean regions in the world.

While a change in average water temperature of 1°C over 30 years or more may not seem huge, it is a significant shift for many sea creatures, which have evolved over thousands of years to live within a relatively narrow temperature range. Thirty years is not enough time for these creatures to evolve new temperature tolerances, so where they can, such creatures will instead move to deeper or more southerly waters in an effort to stay within their ideal temperature range. Shifts in the distribution of populations of a wide range of sea creatures, especially more-mobile fish, have now been observed in many countries where good historical information is available. Such data on shifting populations of sea creatures in response to warming waters are not readily available in New Zealand. However, there are increasing accounts of fish species which normally inhabit warmer waters in the north of New Zealand, such as kingfish and snapper, being seen more frequently in more-southern parts of the country, consistent with them tracking the warming waters into the southern regions of the country.

From overseas studies we know that some marine species have been unable to adjust their distributions due to their reliance on restricted habitats or food sources. In these cases, populations have gone into decline as their ocean system has changed. In other cases, marine species themselves can tolerate the changing temperature conditions, but go into decline because other species they rely on for food or shelter are unable to cope with the changes and decline first, taking with them these associated and dependent species. A similar set of responses is expected for New Zealand's marine species, especially those with existing restricted distributions or specific habitat needs.

Ongoing changes in the flows of ocean currents will also influence the mixing of nutrient-rich cold waters from the southern regions with the warmer and less -nutrient-rich waters to the north.

Increasing seawater temperatures will have other effects on marine species. For example, snapper grow faster when seawater temperatures are higher than normal, and more juveniles survive in summers with warmer water temperatures. Therefore snapper populations can be expected to thrive in warming coastal waters, as long as their nursery habitats can support them and they are not overfished.

Besides changes in average seawater temperatures, our changing climate is also increasing the variability in seawater temperatures. This is best evidenced by the increasing occurrence of 'ocean heatwaves' or periods of exceptionally high surface seawater temperatures. For example, in the summer of 2017/18 the entire New Zealand region experienced an extended marine heatwave, which drove sea-surface temperatures in the eastern Tasman Sea off the western seaboard of New Zealand to over 3.5°C higher than normal. The increasing frequency of these extreme temperature events will also cause challenges for the narrow temperature range of many marine organisms, and may further restrict the geographic ranges of these species.

Ongoing changes in the flows of ocean currents will also influence the mixing of nutrient-rich cold waters from the southern regions with the warmer and less-nutrient-rich waters to the north. These changes are expected to reduce the overall amount of sunlight energy converted by floating plant-like organisms, known as phytoplankton, into the huge quantities of energy-rich biomass that form the critical starting point for ocean food chains. Consequently, this reduced primary productivity is expected to cause widespread disruption throughout many of our ocean ecosystems.

The ecological and economic ramifications of warming ocean waters around New Zealand are major, mostly due to the vulnerability of marine ecosystems and species to the rapidly warming conditions. In many coastal areas, these impacts are expected to be much more severe, exacerbated by existing stressors on the marine environment including fishing pressure and seabed disturbance from trawling, as well as land-based discharge of sediments, nutrients and pollutants. Important marine habitats and species are expected to decline or shift their distribution, and new marine species are expected to arrive and establish in the changed conditions, some of which will have negative consequences. For example, the relatively recent arrival and establishment of two marine pest species in New Zealand, a fan worm and a sea squirt, has been estimated to cost over $1 million a year in lost aquaculture (sea-farming) production and through the implementation of additional control measures.

Out of all economic activities that rely on the ocean, seafood production is expected to be the most intensively affected by the ecological changes in New Zealand's ocean conditions due to climate change. The local seafood industry currently harvests over 700,000 tonnes of seafood a year, from both fishing and aquaculture, and the industry exports around 270,000 tonnes of product, earning over $1.8 billion. Seafood is New Zealand's fifth-largest export by value, and forms an important part of the country's international reputation as a source of premium food.

Although the commercial fishing industry harvests more than 160 different marine species, the majority of the value comes from a very small number of highly prized or abundant schooling species, many of which are already showing signs of vulnerability to changing ocean conditions, e.g. rock lobster, hoki and pāuā. For example, the rock lobster populations that form the basis of some of the most valuable commercial fisheries around the world, including New Zealand, have had recent major downturns in their breeding success, thought to be due to increased water temperatures affecting larval survival. Huge fluctuations in recruitment of youngsters into our single most valuable commercial fishery, hoki, also appear to be linked in some complex way to shifting climatic conditions. Without an understanding of how these processes work and their impact on fish populations, it becomes difficult to set appropriate catch limits.

Aquaculture is not immune from the effects of warming waters either, with very high numbers of farmed salmon and greenshell mussels at some farms dying during recent summer heatwaves. These marine heatwaves were also thought to have triggered unprecedented toxic algal blooms, which resulted in extended harvesting shutdowns in some key mussel-farming areas of the country. Sharp declines in wild sources of mussel-seed that underpin the valuable greenshell aquaculture industry have also been associated with climatic changes, reducing overall industry production.

In contrast to these negative economic impacts, new opportunities are likely to emerge from some important species benefiting from warming oceans around New Zealand. For example, expanding snapper populations may provide for increased harvesting, and populations of valuable mud crabs may become established in northern areas, rather than just appearing as infrequent visitors from Australia.

Overall, increasing variability and uncertainty in ocean conditions, and the corresponding response to the changing conditions by marine organisms, will make it more difficult to effectively manage important marine species, especially species which are the basis of seafood supplies. Changing ocean conditions will also reduce the resilience of marine ecosystems and their inhabitants, making them much more vulnerable to overexploitation. This is an important issue in New Zealand, where fisheries management strives to maximise ongoing harvest, frequently resulting in the overexploitation of fish stocks. A more conservative approach to managing these stocks is warranted, given the increasing uncertainty around how fish populations will respond to environmental variability.

Changing ocean conditions will also reduce the resilience of marine ecosystems and their inhabitants, making them much more vulnerable to overexploitation.

In addition to the impact on the commercial exploitation of our oceans, we can also expect to see marked changes in marine biodiversity. For example, New Zealand has the highest diversity of seabirds of any country in the world, supporting 87 species which live and feed in our waters. Of these, an astonishing 37 species are endemic and breed nowhere else in the world. Regardless of the high mobility of many seabird species, there are clear indications that many of these species are being affected by changes in ocean ecosystems due to climate change, especially in the distribution and abundance of their key food species.

Rising sea level

Over the past 60 years, sea levels around New Zealand have risen on average by over 2 mm per year, i.e. a total of about 0.15 m. If the current trajectory of climate change continues, without effective global reduction in greenhouse gas emissions, these sea levels are expected to increase by a further 0.21 m by 2040, and 0.67 m by 2090. A 0.8 m rise in sea level by the end of this century would mean that normal high tides would almost daily reach the current one-in-100-year high-tide level. This will push saltwater into freshwater habitats along coastal margins, making existing freshwater streams, rivers, wetlands and aquifers saline for the first time, a process known as salinisation.

In addition to sea-level rise, the increased frequency and intensity of storms will mean greater potential for storm-surge events on our coasts, which will also push seawater further inland. Salinisation destroys freshwater habitats, making them uninhabitable for their usual inhabitants and disrupting their ecosystem functions. As a result, local extinctions of native freshwater species and shifts in their distribution inland are expected to occur. For example, lowland freshwater spawning areas for whitebait are expected to be lost or displaced if they are further salinised.

Increasingly saline conditions also disrupt the natural cycle of nutrients in freshwater systems, typically resulting in a decreased capacity to remove nitrogen and store carbon, while greatly increasing the production of some toxic compounds. The reduced ability of lowland freshwater habitats to arrest nutrient and sediment runoff from the land will place additional pressures on the near-shore ecosystems surrounding New Zealand. These are typically already stressed by excessive sedimentation and nutrient runoff resulting from poor land management, modification by damming, reclamation and causeway construction, and overharvesting of marine species. Consequently, the shallow coastal margins and estuaries that many of us know and love are the marine ecosystems that will be most affected by climate change in New Zealand.

Shallow coastal ecosystems are hugely valuable ecologically, economically and culturally. Some of the most productive habitats on Earth, such as estuaries, shallow harbours, kelp (seaweed) forests and shellfish beds, are located in coastal environments, and have been shown in other parts of the world to be highly vulnerable to the effects of climate change. For example, coastal kelp habitats are extremely productive, capturing sunlight energy through photosynthesis and making it widely available as seaweed tissue for a wide variety of marine animals to feed upon. Large brown seaweeds typically grow in dense groups or extensive stands on the coast, which are often referred to as 'forests'. Like a forest on land, these dense stands create a unique habitat with a complex physical structure which makes them highly biodiverse and suitable as nursery habitats for many fish and other species. Likewise, shallow estuaries are highly productive because plant-like microbes living there use photosynthesis to capture both the readily available nutrients washed from the land and the abundant sunlight shining on intertidal flats and shallow waters. Our estuaries and harbours have

Rising sea levels will result in the disappearance of many intertidal areas in New Zealand

billions of these productive microbes that drive these ecosystems with their captured nutrients combined with the sun's energy. These shallow-water habitats also provide important nursery areas for some of our favoured fishes, as well as the food resources that fuel the migration of our many shorebirds.

Rising sea levels will result in the disappearance of many intertidal areas in New Zealand, important because of their productivity and biodiversity and the variety of ecosystem services they provide, including the reduction of excessive nutrient runoff from land, protection of coastal margins from wave surge, and provision of nursery habitats for mobile coastal species, especially fish. New Zealand shorebirds, which live on coastal fringes and frequently feed in the intertidal shoreline, will be heavily impacted. Today more than a quarter (four out of fourteen) of New Zealand's resident native shorebird species are already at risk of extinction from the existing pressures on these important coastal margins.

As an island nation, the coastal margins are of particular social and cultural significance to New Zealanders. However, ongoing sea-level rise can be expected to diminish or submerge many currently inhabited coastal areas of importance, such as popular beaches, marae, urupā, homes and holiday dwellings that are sited on low-lying coastal land.

Rising sea levels will also reduce freshwater availability for some coastal aquifers around New Zealand, due to changes in the water table that may also result in saltwater intrusion into aquifers, greatly reducing their usefulness.

Coastal engineering, such as building protective coastal groynes and barriers, can be used to protect coastal features and infrastructure. However, they are very expensive to build and maintain. Further, the continuing increases in sea levels will also continue to outdate these structures, so that they will need to be either abandoned or rebuilt to meet higher sea levels. Ultimately, planning for avoiding placing infrastructure in coastal margins, or re-siting infrastructure likely to be affected by rising sea level, will be the most effective response.

Changing seawater chemistry

As a result of the higher amount of carbon dioxide in the air, more of this gas is dissolving into the oceans, which alters the chemistry of seawater, including making it more acidic. Although variable with time and place, the overall increasing acidity of coastal waters has already been detected in regular sampling off the Otago coast that has been carried out since 1998.

FIGURE 1. Spot the difference between New Zealand today and in 2100 with the increasing effects of climate change. Key differences in the future picture are a higher snow line, heavier rainfall events, much more of the land forested, the shoreline eroded with loss of beaches and dunes, loss of shorebird and seabird species, greater storm waves, loss of kelp habitats, loss of coastal properties, and opportunities for formed saltmarshes on inundated low-lying land.

Acidity is measured on the pH scale, and projections for New Zealand's coastal waters indicate a decrease in the pH of surface seawater (meaning it is more acidic) from a current average pH of 8.09 to 7.95 by 2050, and 7.75 by 2100. While these changes seem small, the pH scale is logarithmic, meaning pH 7 is 10 times more acidic than pH 8. Hence, the predicted pH of seawater by 2100 is equivalent to more than doubling the concentration of hydrogen ions associated with the increased acidity. This predicted pH of seawater in 2100 is thought to be the lowest pH (i.e. the most acidic) in the past 25 million years. Furthermore, the change in pH this century will be the fastest rate of change in the pH of seawater ever to have occurred.

Increasingly acidic seawater creates problems for a broad range of marine creatures which build chalky external skeletons or shells for their protection. This includes sea snails, kina, sea stars, rock lobsters, crabs, mussels, pipi, pāua and corals. It also includes some microscopic organisms which play important ecological roles in ocean productivity and marine food webs. Studies of these various marine creatures suggest that they are highly variable in their response to increasingly acidic seawater, with some seemingly unaffected while others become stressed or unable to grow new shell. A common feature is that early life-history stages, especially larvae and early juveniles, are less likely to survive in more-acidic seawater. Research also suggests that additional stress on marine creatures caused by living in more-acidic seawater will be compounded by other climate change stressors, such as warming ocean temperatures.

Ocean acidification also alters the overall chemistry of seawater, and has been shown to increase the accumulation of the heavy metal cadmium in certain shellfish. In some coastal waters of New Zealand, such as Foveaux Strait, cadmium levels in shellfish are already relatively high in terms of being safe for human consumption. With increasing ocean acidification, there may also need to be greater vigilance of cadmium accumulation in edible shellfish such as oysters, mussels and surf clams, sourced from those coastal waters which already have naturally elevated levels of cadmium.

Overall, it is likely that ocean acidification will also contribute to the economic, cultural and ecological impacts of climate change along New Zealand's coasts. Edible shellfish, which have important economic, cultural and ecological roles, are likely to be most heavily affected. Greenshell mussel aquaculture production, currently valued at over $350 million a year, is particularly vulnerable, but so too are other coastal shellfish species, such as pipi, pāua and kina, that are not only an economically important resource but also a taonga, being highly valued for recreational and cultural reasons.

Changes in weather patterns

Over the past 100 years, the New Zealand climate on land has warmed by around 1°C on average. This warming is expected to continue to accelerate, although the extent of this will depend on global progress on reducing the ongoing release of greenhouse gases. A consequence of the warming New Zealand climate will be a reduction in rainfall and increased drying of the land, which will have impacts on freshwater and terrestrial ecosystems that rely on rainfall. For example, droughts are expected to increase in frequency and severity, especially on the eastern sides of the Southern Alps and in Northland, but less so in Taranaki-Manawatū, the West Coast and Southland. The weather is also expected to become more variable, with extreme weather

Flooding events are more likely as a result of increased occurrence of extreme rainfall events.

events such as heatwaves, storms and heavy rainfall becoming more frequent and intense. Flooding events are more likely as a result of increased occurrence of extreme rainfall events.

These marked changes in weather patterns will have a

significant impact on both freshwater and land ecosystems, which will be disrupted by the rapid changes in climate, creating difficulties for protecting native biodiversity, such as forests and birdlife. The changes in freshwater flow regimes will have significant impacts on streams and rivers, as well as the receiving waters — lakes, wetlands, lagoons and estuaries. Increased drought in many parts of the country will dry out some streams and rivers, leading to loss of their aquatic inhabitants and ecosystem functioning, such as the cycling of nutrients and the production of aquatic insects which provide food for fish and birdlife. Increased extreme rainfall events will denude streams and rivers of their native inhabitants, erode riverbanks and remove riparian vegetation. Riverbanks and their overhanging vegetation are particularly important for providing food and shelter for many native freshwater fish species. For example, streamside vegetation provides important spawning habitat for the native fish inanga, the migrating young of which are a major component of whitebait harvests. High river flows resulting from extreme rainfall events will also affect the water quality in lakes and rivers by increasing their loading of nutrients, sediment and faecal contamination from land runoff. Increased contamination of freshwater with waterborne micro-organisms that cause illness in humans is expected to elevate public health risks for drinking-water supplies and swimming in contaminated water.

Land-based primary production activities, such as pastoral farming, will also need to deal with changed climatic conditions, although some aspects may be beneficial. For example, warmer and longer summer and spring seasons are likely to extend pasture and forest growth where rainfall permits. It is the changes in rainfall that will require the most attention in the farming sector, requiring a variety of mitigation measures, such as increased investment in water storage and irrigation systems and greater stored seasonal-feed supplies in those areas of the country that will become drought-prone. Demand for water in these areas will increase markedly, creating competition with other water uses, such as urban supplies, hydroelectric generation and recreation, as well as conservation. Conflict over water use has recently received media attention, as extreme drought conditions in the Auckland region create demands to extract more water from the Waikato River — a resource that is already intensively drawn on for agriculture, industry and cooling a major thermal power station.

Fortunately, increased rainfall in parts of the South Island is expected to create a slightly larger overall supply of sustainable hydroelectric electricity for New Zealand, making up for any losses from other parts of the country which have declining rainfall.

A warming climate in New Zealand will have a significant impact on the ecology of lakes, through reducing the mixing of the water column. The mixing of water within lakes is particularly important for circulating oxygen-rich surface waters to maintain the organisms living in the deeper waters of lakes. Many fish species, as well as bottom-dwelling species including kōura (freshwater crayfish) and kākahi (freshwater mussels), cannot survive in oxygen-depleted conditions. The water temperature within lakes is also regulated by mixing processes, with warmed surface waters being cooled through circulation of cooler waters at depth, and vice versa.

Mixing of the water column also helps to redistribute nutrients within lakes, preventing zones of nutrient depletion or accumulation, which can otherwise trigger blooms of algae. The presence of algal blooms in the surface waters of lakes produces debris which sinks into deeper water, where its breakdown depletes the available oxygen and reduces water clarity. In turn, the reduced water clarity restricts sunlight penetration into lake waters and reduces the extent of native aquatic plants, which are important for generating oxygen into the water while also providing food and habitat for many lake-dwelling organisms.

The greatest impact from the decrease in mixing of New Zealand's lakes resulting from climate warming is expected to occur in the many shallower and coastal lakes. In particular, more-frequent and longer periods of reduced mixing will reduce the habitat quality of the deeper waters for resident native species.

Conclusion

The most obvious effects of climate change on aquatic ecosystems are already beginning to be seen in New Zealand, such as increases in summertime mortalities of shellfish beds and heavy rainfall events carrying large quantities of forestry debris down rivers and onto the coast to cover beaches. However, many other effects of climate change on aquatic ecosystems will not be as obvious — or, more likely, changes will occur but it will be difficult to link these changes to any specific aspect of climate change.

Establishing causal links between shifts in aquatic environments and climate change is challenging for several reasons. First, multiple environmental changes are being generated through different processes resulting from climate change, so that organisms are likely to be affected by the cumulative stressors resulting from a variety of processes. For example, shellfish species living in shallow coastal waters, such as mussels, pipi and cockles, will experience increased water temperatures, more-acidic seawater, greater storm disturbance and greater fluctuations in salinity and seawater quality, resulting from increased land runoff from increasingly heavy rainfall events. Hence, attempting to isolate a single cause of diminishing populations of these coastal shellfish species would be futile because of the cumulative effects they are experiencing.

Second, aquatic ecosystems by their nature are highly interconnected, making it extremely difficult to trace the cause of any observed changes. Food chains in aquatic ecosystems are complex, and swiftly and efficiently pass along energy and nutrients, meaning that any disturbances can result in rapid and widespread changes in the ecosystem. Changes also tend to be magnified for those species higher up the food chain, such as hāpuka, sharks, dolphins and large seabirds. Consequently, the existing perilous state of many of New Zealand's seabirds and shorebirds is very likely to be exacerbated by climate change effects on our oceans.

Third, marine and estuarine ecosystems frequently operate with strong feedback processes, so that even small perturbations may produce dramatic changes. For example, kelp forests provide nursery habitats for rock lobsters, which in turn grow up to become a major predator for kina, thereby controlling the size of kina populations and preventing them from grazing out the kelp forest. Losing kelp forests because of increasing seawater temperatures, greater sedimentation or the introduction of new warm-water species which feed on kelp will result in reduced numbers of young rock lobsters, which in turn will leave kina populations without predators, ultimately resulting in further loss of kelp forests.

Managing aquatic ecosystems in the face of increased natural variability and uncertainty generated by climate change requires a much more cautious and adaptable approach than is normally the case at present. Greater leeway will need to be built into catch limits for commercial fisheries to accommodate unexpected shifts in fish populations, and the fishing industry needs to be prepared to detect and move quickly to adjust

Some changes to aquatic ecosystems in New Zealand appear certain and can be planned for more easily.

to such changes. Monitoring will become increasingly important for enabling more responsive and effective management regimes. Consideration should also be given to greatly relieving fishing pressure on species that are more likely to be affected by ecological shifts, such as those higher up the food chain, species operating at the limit of their thermal range, those subjected to multiple stressors and those with strong localised habitat associations.

Some changes to aquatic ecosystems in New Zealand appear certain and can be planned for more easily. For example, increasing summer seawater temperatures in the Marlborough Sounds are making it more difficult to farm salmon in the region, because the temperature exceeds the upper tolerance of the salmon. Plans can be put in place to breed more-temperature-tolerant salmon, switch to farming other fish species better suited to warmer water, or move the salmon farms to locations with colder water. Some of these plans are now being implemented as a response to the anticipated environmental changes. Another certain change that can be planned for is the anticipated effects of rising sea level on coastal infrastructure.

Reduced rainfall in many parts of New Zealand, combined with increasing temperatures, will place greater demands on the remaining freshwater resources for irrigation, watering stock, and domestic and industrial use. Meanwhile, natural freshwater habitats will be stressed due to decreased water flow during times of drought. Outside of droughts, more-frequent heavy rainfall events will lead to flooding and erosion of natural watercourses. Managing freshwater resources effectively will become increasingly challenging, due to the need to resolve the competing demands during times of drought while also limiting the damage resulting from extreme rainfall events.

Given the ongoing increases in global greenhouse gas emissions, further climate change effects on our aquatic ecosystems are already locked in for the rest of this century and beyond, unless a revolutionary global solution is discovered and implemented immediately. New Zealand has followed the same trend of the international community by continuing to increase its greenhouse gas emissions in recent years. Overall, New Zealand's emissions have increased by around 25 per cent from 1990 levels. Rapidly reducing our overall greenhouse emissions is a priority at all levels of New Zealand society, from individuals to central government. As a nation we can also encourage and assist other nations to follow suit as part of an international response to the global climate crisis.

Other than addressing the root cause of climate change, it is difficult for us to undertake measures that will directly prevent the impacts it will have on our aquatic environments. However, given that climate change impacts act in concert with existing stressors on aquatic environments, we can do more to reduce these existing stressors, as we have greater control over them. The runoff of nutrients and sediment from the land can be managed through greatly improving catchment management and more carefully controlling fertiliser application. The more impactful fish-harvesting methods, such as seabed trawling and dredging, can be greatly reduced or eliminated. We can also create more marine protected areas to act as biodiversity repositories and help to continue restocking locations affected by climate change processes.

The need to arrest climate change will alter the use of areas of land and sea in New Zealand, which will create opportunities to greatly improve outcomes for our aquatic environments. A desire to increase carbon sequestration through revegetating land with trees and wetlands is an excellent opportunity to also target the margins of lakes and waterways to help reduce sediment and nutrient runoff into these ecosystems. Likewise, inundated areas of low-lying coastal land, especially previously drained pastoral land, provide an opportunity to create new saltmarsh habitat for protecting biodiversity, while also creating important new carbon sinks.

Increasing aquaculture activity is a means of producing high-quality food and industrial resources with vastly lower greenhouse gas emissions compared with our existing pastoral farming of cows and sheep. Pastoral farming relies heavily on potentially polluting nutrient inputs and results in high greenhouse gas emissions from both the digestion processes of the grazing animals and the interaction of their urine and dung with the pasture. Greenhouse gases from agriculture currently make up around half of our total national emissions, releasing the equivalent of nearly 40 million tonnes of carbon dioxide in 2018. Rapid reductions in emissions from pastoral production are required, but are technically, politically and economically challenging to deliver.

The development of aquaculture as an alternative to pastoral farming could help to reduce the pain of the transition from this traditional industry. For example, the development of aquaculture of seaweeds in this country could be used to produce high-quality food and industrial products while also boosting marine carbon sequestration, and stimulating the productivity of coastal ecosystems.

Delivering these kinds of changes in New Zealand quickly will rely on wider public and political recognition of the urgency of the problem, and the need for us to make substantial changes to our economy and our way of life to help protect the aquatic environments we prize.

PART 2

The Issues

From threat to opportunity: Climate change and health in Aotearoa New Zealand

KERA SHERWOOD-O'REGAN
Impact Director, Activate Agency

DR RHYS JONES
Te Kupenga Hauora Māori, University of Auckland

In 2009 one of the world's most esteemed medical journals, *The Lancet*, declared climate change the 'biggest global health threat of the 21st century' (Costello et al., 2009). The report confirmed the health challenges ahead, including: direct impacts of a changed environment due to increased temperatures and climate disasters; biological impacts like the changing patterns of infectious diseases; and socially mediated impacts which emerge as consequences of these changes, like food shortages and conflict. Six years later, the same journal declared climate change the most important health opportunity of the century, highlighting the massive potential health gains of ambitious and effective climate action (Watts et al., 2015). So how do we ensure that our collective action on climate change helps to build a healthier world, rather than entrench climate threats to our wellbeing, and how are climate change and health related anyway?

In public health we are frequently concerned with the issue of inequity, or the differences in health and wellbeing that come about through unequal access to the prerequisites of a healthy life. In Aotearoa, health inequity holds a mirror to existing socioeconomic divisions in our society reinforced by structural and generational long-term effects of oppression, including colonisation and capitalism. Inequities persist in our society across income, education, employment and housing — all factors known in the health sector as 'determinants of health' (Dahlgreen & Whitehead, 1991). It comes as no surprise that certain communities and segments of our population tend to experience better health outcomes, while others experience greater ill-health and disease — and these determinants help us understand why.

The relationship between health and wealth is key to understanding the role that determinants of health play in everyday life. Consider just one of the determinants: employment. Whereas so-called 'knowledge workers' are typically more able to work from home, or have more-flexible sick leave policies, the picture looks different for so-called working class roles. If your employer has limited provisions for sick leave, this prevents you from taking time off when you're sick or requires an expensive doctor's note. You are therefore more likely to continue working through illness, and could — as we've seen in 2020 — spread communicable diseases like colds, flus and viruses to your colleagues and others around you. Similarly, if a lower income means you can't afford to buy fresh and healthy foods for a balanced diet, or if it's unsafe to exercise outdoors in your neighbourhood, it follows that you're more likely to experience health complications, such as heart disease, than those who can afford expensive fresh fruits, vegetables and other healthy nutrition, as well as an expensive gym pass to exercise in a convenient and safe environment.

What we so often conceive of as healthy behavioural 'choices' for some people are not really choices at all, and the more marginalisation people experience, the fewer options are available in the first place. These same communities are likely to experience the impacts of climate change first and worst. They may, for instance, have fewer financial and social resources to access climate-safe housing, or fewer mental health services to cope with climate trauma like the loss of culturally significant areas to coastal erosion, flooding and fires. As climate change worsens, health inequities between poorer communities and the most privileged are likely to be magnified, despite the fact that the marginalised and oppressed in society contribute least to climate-damaging emissions.

This double burden is not coincidental, but rather highlights that the systems responsible for poor health are the same systems that produce climate change. Indeed, the largest contributors to carbon emissions and environmental destruction — think mining, drilling, logging, intensive agriculture, and other resource extraction — in many cases represent the same relentless demand for productivity at work in our own lives, widening inequalities in health. This capitalist system with an imperative to growth at all costs, including costs to workers' health and to the climate, furthermore relies on whenua, or land, that has been forcefully taken from Indigenous Peoples through ongoing systems of violence and oppression that continue to feed back into climate destruction and the erosion of our health. Colonialism, with a capitalist economy in tow, has radically disrupted and subjugated Indigenous societies and systems worldwide, and has resulted in an economic system that is fundamentally out of step with the requirements of our environment and healthy human habitation.

Alongside the damning projections for widening health inequity if our society continues its current climate trajectory, *The Lancet* also highlighted the significant positive potential for climate action to improve the health and wellbeing of our communities. An increase in active modes of transport, like walking and cycling, improves the health of our climate by reducing carbon emissions but also improves, for example, cardiovascular fitness. If we reduce the burning of climate-damaging fossil fuels, we can improve overall air pollution which contributes to asthma and other respiratory diseases. With 80 per cent of the world's biodiversity existing in Indigenously managed lands (United Nations Environment Programme, 2017), taking active steps to restore ownership and management of lands to Indigenous Peoples can improve the cultural wellbeing of those communities as well as kickstart the reforestation and restoration of lands necessary to combat the worst effects of climate change.

As members of OraTaiao: The New Zealand Climate and Health Council, the authors share a vision for 'Healthy People and a Healthy Planet'. It is clear that in order to reach this lofty goal, we must not only lean on western medical and scientific knowledge but also draw on the understandings of climate change and wellbeing held by tangata whenua and other frontline communities who have their own critically important perspectives and valuable expertise.

In this chapter we will dive into the health impacts of our changing climate, addressing how they are spread unequally across our society, with a focus on Aotearoa New Zealand. We will consider how action to prevent climate change can not only prevent some of these adverse health impacts but also positively affect existing health issues across our communities to create a healthier society as a whole. Finally, we will assess what actions have already been taken both within our health sector and outside of it, and consider what is needed to ensure a climate-safe approach to health for all.

PART ONE: Health impacts

Climate change affects human health in many ways, ranging from injuries in climate disasters to affecting the underlying conditions of society that support or impair our wellbeing. To conceptualise the depth and breadth of these impacts, it can be helpful to categorise these effects into direct impacts, biologically mediated impacts and socially mediated impacts, although there is of course overlap between all three.

Direct impacts occur in direct response to climate change, and include injuries and trauma from climate disasters and illnesses related to extreme heat. Biologically mediated impacts are those that occur through a biological or natural mechanism or carrier, like infectious diseases caused by pathogenic bacteria, or viruses that thrive in new climates or are carried by animals and insects like mosquitoes or ticks. Socially mediated impacts occur as a consequence of human responses to climate change and cover a wide range of impacts — from increased conflict or political strife to the trauma of seeking asylum when ancestral homelands are no longer habitable, and reduced access to health services due to changing economic and political situations.

While many of the impacts that are outlined below can be alarming, it is important to recognise that all occur as a matter of degree. The more robust and ambitious action we take, the more we can limit warming, the less severe the impacts are likely to be. Every fraction of a degree of warming we can prevent is a reduction in our overall risk, and action towards keeping safe our most affected communities.

DIRECT IMPACTS
Floods and storms

Climate change is associated with increased frequency and severity of extreme weather events such as floods and storms. Floods are the most common of all natural climate disasters, and contribute significantly to the overall health costs of our changing climate. Cases of river flooding have been rising globally, and within Aotearoa we are also becoming increasingly accustomed to flood events of increasing severity.

Immediate impacts of flooding include death by drowning, hypothermia, or injuries such as from destabilised debris. However, the health impacts do not evaporate with receding flood waters — and floods are rarely cleaned up quickly. Ongoing dampness and water damage can create the perfect environment for fungi and mould growth, which contribute particularly to respiratory conditions and the exacerbation of existing asthma and allergies. Flooding also carries an increased risk of waterborne infectious diseases. This includes diarrhoea, leptospirosis, cholera and other diseases carried by vectors that breed in stagnant water, such as mosquitoes (Smith et al., 2014).

Flood events also cause significant harm to people's mental health, including depression, anxiety and psychological distress, as well as exacerbating existing mental distress (Smith et al., 2014). In a UK study, mental health symptoms were found to be two to five times greater for people who had experienced flood water in their homes, and in follow-up studies of victims of Hurricane Katrina in New Orleans, hurricane-related mental health symptoms were experienced by many residents two years after the event (Smith et al., 2014). This is not surprising given that extreme weather events can be uniquely destructive not only in the immediate occurrence but also in the changes they cause to people's social and economic circumstances — all of which are compounded in populations already experiencing structural oppression and for those who are already geographically isolated.

Fires

Wildfires are projected to increase with rising temperatures, often precipitated by droughts and heatwaves. As we witness year on year, most recently in California and Australia, fire events cause catastrophic damage with severe health implications for those in the immediate vicinity, and also affect the health of communities far from the fire site. Immediate impacts such as physical trauma from destabilised structures, burns, heat exhaustion, and death from smoke inhalation are cause for serious alarm (Smith et al., 2014). Yet again, these are particularly exacerbated for communities who are already disadvantaged, with obvious links to the risks of inadequate housing, geographic isolation and lack of financial resources. Elderly and disabled people, who are likely to be less mobile, may also have less ability to escape natural disasters such as fire and additionally may be more susceptible to heat-related death.

Outside of the immediate perimeter, fires can also have a significant impact on the health and wellbeing of communities far away from the site, associated with the toxic chemicals that are projected into the air, as we witnessed with asthma-inducing ash clouds in Aotearoa during the January 2020 Australian wildfires (Xing et al., 2016; Walter et al., 2020).

Heat

Perhaps the most obvious direct impact of climate change on health are those effects related to high temperatures, such as heat exhaustion, heatstroke and related conditions. Heat exhaustion occurs when human body temperature reaches above 38°C and causes a variety of symptoms, such as feeling lethargic and generally unwell, as well as slow thinking and difficulty concentrating. More seriously, heatstroke occurs when body temperatures reach over 40.6°C, with extremely dangerous effects on body systems. These can include organ damage, loss of consciousness, altered mental state and circulatory collapse.

Heatstroke is a medical emergency, deadly if untreated (Al Mahri & Bouchama, 2018).

The impacts of heat-related health conditions are not distributed evenly in global or local populations. On the whole, older people and those with pre-existing health conditions are much more likely to experience severe impacts of extreme temperatures. Key factors that reduce heat exhaustion and heatstroke overlap with adequate standards of housing, including good ventilation like air conditioning, as well as how humid the immediate surroundings are and how much heat is radiated from a building. It follows, then, that extreme temperatures have more adverse effects on people of lower socioeconomic status who may have less access to these things, and who are already disproportionately affected by chronic health conditions that reduce their ability to withstand extremes of temperature.

Many who have experienced heatstroke and been successfully treated still experience long-term symptoms, including neurological disability (Al Mahri & Bouchama, 2018). It is important to recognise that these impacts relate specifically to body temperatures, which do not necessarily have a direct relationship to the ambient temperature outside. Treatment of heatstroke involves rapidly cooling the body to reduce core body temperature, which can be done in hospital such as by immersing the patient in an ice-water bath. This requires multiple medical staff and significant hospital resources (Al Mahri & Bouchama, 2018), which could pose difficulties for treatment if widespread and more severe heatwaves caused patient demand to outstrip the capacities of local hospitals, especially in rural or low-income areas.

BIOLOGICALLY MEDIATED IMPACTS
Allergens and allergic diseases

Carbon dioxide is a key resource for plant development, so with increasing concentration in the atmosphere due to fossil fuel emissions, deforestation and other high-emissions practices, as well as the warmer temperatures predicted with climate change, many plants are likely to increase their growth and reproduction thanks to this plentiful fuel (Katelaris & Beggs, 2018). What this means for our health is higher quantities of pollen produced throughout a longer pollen season, which can increase the severity of symptoms for those who suffer from pollen allergies. Symptoms include hay fever (sneezing, and a runny or blocked nose), wheezing and an itchy nose, throat or eyes.

Another feature of our changing climate that will have serious consequences for allergic diseases is flooding. Flooding is set to become drastically increased in Aotearoa and among our Pacific neighbours. One of the consequent impacts of flooding and humidity is dampness in homes, creating a perfect environment for the growth of moulds and fungi. Increases in allergens have an exaggerated impact on those who already live with respiratory conditions — a major feature of the New Zealand health landscape. Children are especially susceptible to many allergic conditions, and the rate of child eczema is also closely associated with humidity (Smith et al., 2014). One in seven children and one in eight adults in New Zealand have asthma which is typically associated with damp housing and overcrowding, and this disproportionately affects our whānau Māori and Pasifika, and those living on lower incomes (HQSCNZ, 2020). Given Aotearoa's track record for allergies and damp housing, widening inequities in respiratory health outcomes is an impact of climate change that we will have to be particularly careful of.

Communicable disease

Changes to our climate play a major role in the distribution of disease-causing bacteria and viruses, known as pathogens. These are commonly associated with tropical regions of the world, but with temperature rises and changing rainfall patterns they are projected in some areas to spread into neighbouring regions. Communicable diseases can be spread between humans directly, and are likely to proliferate in environments created in the aftermath of natural disasters, or indirectly through contact with contaminated food, water, body fluids or surfaces. Communicable diseases of particular cause for concern are those that affect our digestive system (Smith et al., 2014).

In Aotearoa, such diseases are affected by extremes of both drought and high rainfall. Drought can affect water supplies, making it harder to keep up hygiene practices like good handwashing. Temperature increases are also likely to influence the growth of pathogenic bacteria and parasitic diseases that cause gastroenteritis — infection of the digestive system that typically results in diarrhoea, vomiting and other unpleasant bowel symptoms. High levels of rainfall can cause drains and reservoirs to overflow, risking contamination of clean drinking water with faecal pathogens.

Giardia is a common parasitic disease affecting the digestive system. New Zealand has some of the highest rates of giardia infection for a wealthy country, and it is commonly caught from streams and lake water which have come into contact with affected animals like farm animals, dogs or possums. When people inadvertently eat or drink something contaminated by this water, it causes diarrhoea, stomach cramps, bloating and other uncomfortable symptoms (MOH, 2020). It is possible that giardia may spread more extensively due to increased rainfall as a result of climate change, as this could wash giardia cysts present in animal faeces into water sources that we use for food, water and recreation.

Some studies in Aotearoa have also demonstrated a possible link between salmonella infections and higher-than-average monthly temperatures (Wilson et al., 2011). Salmonella bacteria live in the guts of humans and birds, and are transmitted to humans through eating contaminated food. We know this disease as the classic 'food poisoning' symptoms of diarrhoea, stomach cramps, nausea and vomiting. It is typically spread by poor hand hygiene prior to preparing or eating food, and is particularly associated with raw chicken and raw or undercooked eggs and dairy products (ARPHS, 2020).

Vector-borne disease

Vector-borne diseases are spread by an intermediary organism known as a vector — most commonly small insects such as mosquitoes and ticks — transmitting viruses, bacteria or other pathogenic microorganisms into the human bloodstream through insect bites. The spread of these diseases is particularly influenced by changing climatic patterns because many insect vectors often have short life cycles that are heavily influenced by temperature and rainfall. As changes in climate can provide new habitable environments for insects, vector-borne diseases that are projected to increase significantly include malaria and dengue fever, both spread by mosquitoes; and tick-borne encephalitis and Lyme disease, transmitted by ticks.

Mosquitoes prefer warmer and more humid climates, and breed by laying their eggs in standing water. When we consider the changes in climate we are likely to see in Aotearoa, we know that certain areas of the country are likely to become warmer and wetter, providing the ideal environment for mosquito breeding. Even in those dry areas where drought is likely to increase, human behaviour can create ideal breeding conditions for disease vectors. Overseas examples demonstrate that during periods of drought, people have a tendency to collect water and keep it stored in containers — the ideal standing-water habitat for mosquitoes to breed in.

Affecting more than 200 million people every year, mostly in tropical climates, malaria is also heavily influenced by socioeconomic conditions and the status of health systems. Even slight warming can risk increased malaria transmission significantly, and in some regions that have previously eradicated malaria, like Greece, the disease has re-emerged in the presence of less-favourable social and economic contexts (Smith et al., 2014). While malaria is unlikely to gain a foothold in Aotearoa in the near future, the potential impact of this deadly disease throughout the Pacific is worth considering, especially across nations where New Zealand has been a colonising force and has a duty of care.

Dengue fever is a viral condition that causes a variety of symptoms including fever, body aches, pain, rash, nausea and vomiting for almost a hundred million people each year, largely in tropical and subtropical regions (Smith et al., 2014). Three-quarters of these people live in the Asia-Pacific region, and dengue fever significantly affects our Pacific neighbours across New Caledonia, Fiji and other Moana nations. The virus is carried by female *Aedes* mosquitoes, which transmit the virus into human skin through their saliva. The virus then replicates inside human immune cells before being carried to the lymph nodes and spreading to the liver, lungs and spleen. More severe cases may progress to dengue haemorrhagic fever, which causes bleeding problems, and dengue shock syndrome, a fatal condition caused by leakage of blood plasma (Khetarpal & Khanna, 2016).

A warmer planet means that not only will more people be exposed to infections such as malaria and dengue fever, but also that these newly exposed populations are less likely to have developed protections against them, putting them at higher risk. In some regions where such communicable diseases are endemic, communities have built up cultural and biological protections against them. In some communities, many people are carriers of a 'faulty' gene that codes for sickle cell anaemia. This gene also provides some protection against malaria development in those who have been infected, meaning that people with the 'faulty' gene are more likely to survive to reproduce and pass the gene on to their children (Luzzatto, 2012). Other protections are cultural, like normalising the use of mosquito nets when sleeping, or building settlements away from areas that are most affected.

Lyme disease is caused by the *Borrelia* bacteria, which is transmitted to humans by tick bite, and causes rashes, headaches, and joint stiffness and pain. Currently, Lyme disease is found throughout Europe and North America and there are no known cases in New Zealand. However, the tick which carries Borrelia has been expanding its territory in North America with warming climates since the 1990s, and ticks that are endemic to New Zealand are thought to be capable of spreading this disease (Ogden et al., 2014).

SOCIALLY MEDIATED IMPACTS

From social and political movements to government policies, from disaster preparedness at home to political tensions around the globe, our access to the resources and services required to stay healthy are determined by 'social determinants of health' (Dahlgren & Whitehead, 1991). These arise from complex webs of human relationships, interactions and structures that make up our systems and societies. Whether we can access healthy kai may depend on where we live or how much money we earn, which is often determined by our access to education or whether an employer deems our race, gender or class desirable. As much as our health in a changing climate is threatened by exposure to floods or fires, so too is it threatened by increasingly inequitable access to social determinants of health, which overlap and intersect.

Whether we can access healthy kai may depend on where we live or how much money we earn, which is often determined by our access to education or whether an employer deems our race, gender or class desirable.

Disruption to education

Education is a key factor for upward mobility within society, enabling people to escape poverty or otherwise improve their socioeconomic status. It also directly affects a person's or community's health by sharing knowledge of effective healthy behaviours and so reducing the need for healthcare. Education is also associated with lower rates of depression, stronger self-concept and self-esteem, and reduced likelihood to work in hazardous jobs (Feinstein et al., 2006). These effects are evident from primary school to later-in-life tertiary or vocational training. But what happens when that education is interrupted or no longer available due to factors like disease outbreaks or climate disasters?

In 2020, the coronavirus pandemic afforded us a glimpse of the impacts of educational disruption in Aotearoa and around the world. The United Nations reported that in April 2020, 1.6 billion children and young people were out of school due to temporary pandemic closures (United Nations, 2020). Aside from the obvious disruption to learning, these closures also represented a loss of childcare for many parents, with flow-on effects for their employment and income. More than 360 million children who routinely rely on school meals needed new sources of nutrition (United Nations, 2020). This phenomenon affects not only lower-income countries but also students in higher-income countries more generally, and here in Aotearoa it has a predictably larger implication for those whānau who already experience socioeconomic marginalisation. When schools in New Zealand had to adapt to online learning during the strict nationwide lockdown, many whānau reported not having access to the devices or internet connection necessary for their children to participate. In order to close this 'digital divide', the government connected 33,000 homes to the internet and provided devices to 25,000 students as of August 2020 (Wiltshire, 2020). While this government intervention targeted students in the lowest-decile schools, some families still did not receive adequate support. Other factors, like not having dedicated space at home for children's learning and having to share limited space with parents and extended family working from home, were widespread and reinforced existing inequalities.

As climate change worsens, we can expect to see similar disruptions to the lives and education of our children, whether from local outbreaks of infectious diseases or from destructive climatic events and heatwaves. Unless we take decisive action to centre equity and justice in our climate and educational policies, these disruptions are likely to reinforce inequalities along socioeconomic and ethnic lines, with those who are already least served by our education system likely to experience greater disruptions and therefore a higher cost to their health in the long term.

Job losses

Employment is another key social determinant of health. More than simply providing the financial resources for people to pay for health services, healthy food and healthy housing, for many New Zealanders our workplaces and jobs can be a significant source of satisfaction and wellbeing. Employment can positively influence how we feel about ourselves, how we feel about our communities and how we see ourselves within the world. Many supportive workplaces also offer direct health benefits such as access to health insurance. A well-designed office might support our physical wellbeing with access to ergonomic furniture and other work tools, healthy catering or food services, health and fitness classes, social groups and — in some workplaces — even offer direct wellbeing allowances for employees to spend on health-promoting activities like gym memberships and eye tests.

Work can also be a significant source of stress, particularly in a society where some sectors have normalised the 'grind culture' of longer and longer hours, often at a poorly designed desk set up in front of a computer screen. More workers still are expected to assume risk in the 'gig economy', which has boomed in a shift to freelancing and independent contracting across sectors, and in a shift to third-party apps transforming food delivery, taxi driving and other frontline service roles into independent contractors with extremely small margins and financial precarity. This work sees many people in Aotearoa working multiple jobs to make ends meet, and doing so in work environments that are woefully unhealthy for their physical and mental health.

For people currently out of work or unable to work full-time or at all, social security is also financially precarious. In 2020, dozens of community organisations highlighted the extreme inadequacies of unemployment, disability and other benefits which do not cover reasonable costs for people to maintain healthy lives in an increasingly expensive society (Franks, 2020). This can be extremely distressing for many, and forces some people on social security support to prioritise ever-increasing rents over essential foods, medicines and other health requirements.

Differences in employment status have been exacerbated by the Covid-19 pandemic, with many low-wage and casual workers having their hours cut to unliveable levels of pay, or putting people out of work entirely. Many had to apply for benefits to pay their essential living costs. A further inequality has come from the New Zealand government's response to the pandemic, with newly unemployed 'knowledge workers' or self-employed people receiving a much higher level of support for a limited time than those people already on benefits prior to Covid-19, or those working lower status jobs. After experiencing changes to employment conditions and instability through the coronavirus pandemic in 2020, we can imagine how climate disasters or disease outbreaks may affect people's employment and wellbeing in the future.

It's also important to consider how society's attempts to prevent and mitigate climate change can create inequitable outcomes if not approached with climate justice in mind. In New Zealand's move to a more climate-safe economy, every industry will have to play its part. However, some will be more affected than others. High-emissions industries like fossil-fuel extraction, energy production and agriculture must undoubtedly transform their practices on a massive scale, and in some cases drastically reduce or shut down production altogether. As we have seen in other chapters, one of the biggest emissions culprits in Aotearoa is the agricultural sector, with methane emissions a particular concern. While these sectors, and their shareholders in particular, have reaped enormous profits over the years and benefited from relaxed regulations that have allowed them to dodge paying the true cost of their pollution, little of this profit has made its way into the pockets of frontline workers. Many communities throughout Aotearoa, rural communities in particular, rely heavily on high-emissions industries, and would experience mass unemployment and financial hardship if they were not assisted in transitioning to more sustainable work, or supported with retraining for other industries, such as in renewable energy.

A changing diet

Adequate nutrition provides our bodies with the fundamental energy and nutrients we require as fuel to live. In Aotearoa there are already existing disparities in communities' access to good nutrition, most notably the cost barrier to healthy food for poorer and predominantly Māori and Pasifika communities. The consequences are higher rates of obesity-related diseases connected to a diet higher in carbohydrates and saturated fats.

A range of social, societal and biological factors that influence healthy nutrition are also affected by climate change. Agricultural production and crop yields have major impacts on the availability and price of food, creating economic and geographic barriers to a healthy lifestyle. As temperatures rise and rain patterns change, scientists predict that there will be corresponding changes in the types, quantity and quality of foods available for consumption. The total food energy available is likely to decrease on a global scale, with a corresponding increase in childhood malnutrition, disability and mortality (Smith et al., 2014).

While New Zealand as a whole doesn't face the same immediate risk of malnutrition that we may associate with low income countries, unequal access to good-quality nutrition is already a major issue in many communities. There is a risk, therefore, that climate change may widen these existing inequalities and a balanced diet may become exclusive to affluent communities, creating major health repercussions for those who are priced out of access to healthy kai.

Cultural wellbeing

Access to one's culture and cultural practices is not only a key human right but also a key determinant of individual and collective wellbeing. In this regard, climate change poses a particular threat to Indigenous Peoples who are already being forced to leave their homelands due to rising sea levels and other climate events. Forced climate migration may cause immense cultural challenges for people, separating them from sites of cultural significance, relations and communities, and meaning that many will have to navigate what it means to be Indigenous and practise one's culture outside of their territories.

Additionally, through coastal erosion, glacial melting and other manifestations of climate change, wāhi tapu, or sacred sites including urupā (cemeteries), sites inhabited by ancestors, and ancestral mountains are already becoming compromised in Aotearoa and around the world. This may become a significant source of cultural trauma, as well as demanding various interventions from communities to prevent and mitigate damage. In Aotearoa we have already seen a growing concern about climate-related depression and anxiety, particularly in young people, and we should expect this to be magnified with climate change for Indigenous Peoples and any whose cultural practices and beliefs are intimately connected to the whenua and moana affected.

> **In Aotearoa we have already seen a growing concern about climate-related depression and anxiety, particularly in young people.**

Additionally, many Indigenous Peoples and communities in the Global South have been advocating for climate action, and protesting against extractive industries in their regions for generations — often at significant costs to their lives, livelihoods and freedom. Land and water protectors globally have faced extreme violence, intimidation and prevention of cultural practices, often sanctioned or facilitated by colonial governments and police forces — as in the case of Standing Rock and action against the Dakota Access

Pipeline; Wet'suwet'en and access against the Coastal Gas Link Pipeline; and, in Aotearoa, the protection of Ihumātao against destruction of sacred whenua and carbon-intensive development. Indigenous protectors in these situations are under immense pressure, and are often forced into traumatic and culturally harmful situations which are likely to have impacts on their wellbeing for years to come. As more and more lands and waters are affected by climate change, it is possible that such protector movements will come under increased pressure, particularly as increasingly urgent climate imperatives bring them into direct conflict with corporate interests.

Global conflict and violence

As climate change limits access to natural resources, and food and water scarcity becomes a feature of many regions around the world, political tensions, such as conflicts for resources, are likely to follow. Conflict results in violence and injury, but also increased mental distress and trauma for societies as a whole. It exacerbates shortages of food, water, medicines and other necessary health resources, disrupts supply chains and destroys infrastructure. Consequences include malnutrition, disruption to health services including essential vaccinations, and increased rates of communicable diseases (Georgetown University Medical Centre, 2020).

While climate projections indicate that New Zealand is not as likely to experience these levels of scarcity and instability, as a relatively wealthy country that has contributed significantly to climate change and political instability we have a responsibility to support people seeking asylum as climate refugees, as well as to do our utmost to slow the effects of climate change. The process of seeking asylum in a new country can be a profoundly traumatic experience, and at the very least can cause major disruptions to people's access to health services, resources and other social determinants of health.

PART TWO: Opportunities

The picture painted so far is undoubtedly cause for concern. It is clear that climate change, and our responses to it, will be a major feature of the health landscape for generations to come, with significant impacts at an individual level, a community level, on our national health systems, and on a global scale when considering the increasingly globalised world we live in, including the risks of communicable disease spread that have been illustrated by the Covid-19 pandemic.

However, while this concern is justified, and urgent action is needed to prevent these impacts on our health and wellbeing, there is a positive, hopeful side to the story. Indeed, tackling climate change has been described as possibly the greatest global health opportunity of the 21st century (Watts et al., 2015). This is because many of the actions required to combat climate change and reduce our carbon emissions are also good for our health and wellbeing. Therefore, addressing the climate crisis is not just essential to reduce the potentially devastating *future* impacts on our health, but well-designed climate action can also help to create a healthier and more equitable society right now. These positive side-effects of climate action are known as 'co-benefits'. There are wide-ranging social, economic, environmental and other co-benefits, due to the fact that actions to reduce greenhouse gas emissions are necessarily multi-sectoral and have impacts at all levels of society. However, even just considering the health co-benefits in isolation provides a compelling argument for strong, urgent actions to tackle climate change.

In this section we'll explore selected climate change mitigation strategies and summarise their potential impacts on population health and health equity. It should be noted that this is by no means an exhaustive list, but the examples do demonstrate how climate action can be a win–win solution to many of the problems facing society.

ACTIVE TRANSPORT

Transport is a major source of emissions, especially in high-income countries like Aotearoa. In 2018 it was responsible for 47 per cent of New Zealand's carbon dioxide emissions, or approximately 20 per cent of total greenhouse gas emissions. Transport emissions have been increasing over time, which is concerning given the need for total emissions to reduce to zero in a relatively short space of time (Statistics NZ, 2021). The transport system also has critical implications for health and equity. Investment in transport infrastructure in Aotearoa has historically favoured private car use over public and active transport, with considerable adverse impacts on New Zealanders' health (Shaw et al., 2018). Private motor vehicles are also responsible for a significant proportion of New Zealand's transport emissions.

Shifting from car-based travel to active transport is associated with a range of co-benefits for health. Active transport refers to walking, cycling and other forms of active mobility such as scootering and skateboarding. It is also important to note that using public transport almost always includes an active component. Active transport builds physical activity into people's daily lives, thereby helping to reduce the risk of cardiovascular disease, diabetes and other chronic conditions (Celis-Morales et al., 2017). It also helps people maintain a healthy body weight and is associated with better overall health outcomes. Reduced motor vehicle use leads to a reduction in urban air pollution, which contributes to improvements in respiratory and cardiovascular health, and a reduction in noise pollution. Moving trips from cars to active modes also lowers the risk of road traffic injury and death, increases social cohesion and improves mental health (Mandic et al., 2019).

Our existing transport system also creates inequitable outcomes, including health outcomes, for different groups in society. For example, Māori and Pacific people have higher rates of road traffic injuries than non-Māori, non-Pacific people. Māori and Pacific communities also experience poorer cardiovascular, respiratory and other health outcomes that are associated with a car-dominated transport system. Shifting trips from private motor vehicles to active transport could help to mitigate these inequities; for example, the health benefits from cycling instead of driving are potentially much greater for Māori and Pacific communities than for other New Zealanders (Lindsay et al., 2011). These co-benefits for health and equity can be realised with Indigenous-led interventions that address specific community needs and realities (Jones et al., 2020).

Examining the co-benefits of different strategies to reduce transport emissions highlights the importance of well-designed climate mitigation. For example, transitioning the vehicle fleet from internal combustion engines to electric vehicles (EVs) is one of the favoured policy goals in Aotearoa. While this might help to address the emissions problem, it leaves all other problems associated with our car-centric transport system unsolved. Policies to incentivise EV use also tend to be anti-equity, as the high up-front cost of EVs is often out of reach for low-income households. As illustrated above, encouraging a shift from private motor vehicles to active modes and public transport has the potential to bring about wide-ranging social, economic, environmental and health benefits as well as contributing to equity. However, it is important to ensure that such strategies do not further marginalise communities who already experience barriers. For example, active and public transport may be inaccessible for disabled people, so policies to restrict car use can limit their access to essential services and amenities.

DIET AND FOOD SYSTEMS

Globally, food systems are a significant source of greenhouse gas emissions, as well as being a major contributor to a range of adverse health outcomes (Swinburn et al., 2019). The rise of intensive and exploitative agricultural practices has seen emissions from food production and land use change (e.g. clearing forests for farmland) increasing worldwide in recent decades. This has been associated with a shift from traditional diets to more processed foods and animal-based products. These dietary trends increase the risk of type 2 diabetes, coronary heart disease, some cancers, and other chronic diseases, with overwhelmingly negative impacts on population health (Tilman & Clark, 2014).

There is strong evidence that plant-based diets can contribute to better nutritional and health outcomes as well as significantly reducing the climate impacts of food systems. For example, it has been estimated that replacing animal sources with plant-based foods in high-income countries could reduce premature mortality by up to 12 per cent and cut greenhouse gas emissions by up to 84 per cent (Springmann et al., 2018). Experts have called for an urgent transformation of our food systems in order to address the inter-related environmental and health crises (Willett et al., 2019).

These challenges and opportunities are particularly acute in Aotearoa New Zealand. Dietary risk factors are responsible for more than 10 per cent of the overall burden of disease (MOH, 2013), and these adverse impacts fall disproportionately on Māori and Pacific populations (MOH, 2018). In terms of the food system's contribution to climate change, agriculture is responsible for 48 per cent of New Zealand's total emissions. Methane from dairy cattle digestive systems has been a major contributor to the increase in our emissions from 1990 to 2018 (MFE, 2020). Agriculture also has broader environmental and health impacts, for example reduction in water quality in rivers (Julian et al., 2017), loss of freshwater biodiversity (Joy et al., 2018) and contamination of drinking water sources (Phiri et al., 2020) leading to increased risk of waterborne diseases.

The good news is that transformation of our agricultural and food systems, together with the associated changes to New Zealanders' diets, can help to tackle all these problems simultaneously. A shift away from animal-based foods, particularly red and processed meats, towards diets rich in whole plant foods (e.g. vegetables, fruits, legumes and whole grains) can bring about substantial benefits for climate and health (Drew et al., 2020).

Food sovereignty movements

Food sovereignty movements emphasise restoring community control of food, a critical step towards achieving a healthy, equitable and climate-friendly food system. Key elements of food sovereignty are included in a definition from the Declaration of the Forum for Food Sovereignty in Nyéléni, Mali in 2007:

> *The right of peoples to healthy and culturally appropriate food*
> *produced through ecologically sound and sustainable methods,*
> *and their right to define their own food and agriculture systems.*
> *It puts the aspirations and needs of those who produce, distribute*
> *and consume food at the heart of food systems and policies rather*
> *than the demands of markets and corporations. (Declaration of*
> *the World Forum for Food Sovereignty, Nyéléni, 2007, para. 3)*[1]

These movements have arisen in response to the dominance of food systems by multinational corporations and government policies that undermine communities' control over the food they produce and consume. For Indigenous Peoples, they are part of a wider struggle for self-determination that involves the dismantling of oppressive colonial systems. In Aotearoa and many other countries, Indigenous Peoples are asserting the right to regain control over food gathering, harvesting and preparation, and are revitalising knowledge about traditional foods (Moeke-Pickering et al., 2015).

Food sovereignty approaches can not only contribute to reducing emissions and improving diet-related health outcomes, but are also associated with a range of other co-benefits. By emphasising locally sourced foods and Indigenous knowledges and practices, environmental impacts are minimised. Such approaches help to build autonomy and resilience, reducing exposure to fluctuations in global food prices and

---

1   https://nyeleni.org/spip.php?article290

making communities less vulnerable to food insecurity. They also tend to enhance community cohesion due to the close relationship between food producers, distributors, vendors and consumers within interdependent local food networks.

Indigenous food sovereignty initiatives can provide even more extensive co-benefits. For example, māra kai (community food gardens on marae) are associated with a wide

In Aotearoa and many other countries, Indigenous peoples are asserting the right to regain control over food gathering, harvesting and preparation, and are revitalising knowledge about traditional foods.

range of potential social, cultural, environmental, economic and health benefits (Hond et al., 2019). They help to build connections with whenua (land), whānau, hapū and iwi, and social cohesion through hands-on collective activities. This can enhance shared identity among Māori communities, closer links to whakapapa and a sense of connection to place that is critical for cultural wellbeing. Māra encourage the use of te reo and tikanga in everyday community life, and provide opportunities for reinvigoration of traditional knowledge and practices related to gardening. For example, participants may learn about native species and how to use the maramataka (Māori lunar calendar) to guide planting, harvesting and other activities.

HEALTHY HOUSING

Substandard housing is a significant cause of ill-health in Aotearoa. Many of our houses are cold, damp and mouldy, as a result of inadequate insulation, heating and ventilation (Howden-Chapman et al., 2005). This poor-quality housing contributes to health problems such as asthma, other respiratory conditions and infectious diseases. These impacts are felt disproportionately by Māori, Pacific and low-income households, with children in these households experiencing a greater burden of disease and poorer health outcomes (New Zealand Human Rights Commission et al., 2016).

Creating healthy, energy-efficient homes highlights the potential for win–win outcomes with clear environmental and health benefits. A study examining the effect of retrofitting New Zealand houses with insulation found significant energy savings and CO_2 emissions reductions through improved energy efficiency. The associated health benefits were also compelling, with fewer hospital admissions for the elderly and a reduction in days off school and work for younger household members (Chapman et al., 2009).

PART THREE: Healthy and just climate action

When we think about solving any issue, from how to manage our busy schedules to how to prevent something as large and ominous as climate change, we start with identifying and understanding the problem. The better we can understand the problem, and the more nuance and context we can understand, the better we can work towards effective solutions — or that's the theory anyway. In the many decades since scientists and communities first noticed changing climate patterns, we have been privileged to have many great minds (including those you've heard from in this book) dedicated to deeply understanding this problem from many different angles — from climate scientists understanding weather patterns and various natural phenomena, to political scientists and economists studying the impacts on our nations and societies, to frontline communities and Indigenous elders who have been able to observe and pass on the impacts of climate change on their lands and native species over generations.

We have never had a better understanding of climate change, how it happens, what causes it, and how it is likely to manifest across the globe. Yet despite all this information, governments around the world, including ours, are still tediously slow in taking any substantive action, and the policies and actions they do take are consistently unambitious and woefully inadequate in addressing the scale or root causes of the problem. Similarly, many of the major health concerns affecting our populations have been endlessly studied and campaigned on for decades, with political promises and mission statements to reduce inequities, without much of a shift in real outcomes for our communities.

So why, when we are inundated with the latest science and cutting-edge information, are we so behind on actually preventing the rising tides of climate change and poor health already affecting our communities?

Much of the delay lies in an unwillingness, from individuals and from governments, to address the underlying causes of climate change — and instead a clear preference for surface-level actions that do little to address the issue but provide some comfort or PR potential in having 'done our part'. It's much less daunting to focus on shuffling imaginary credits for metric tonnes of CO^2 or not driving to work one day a week than it is to interrogate the underlying behaviours, beliefs and systems that underpin our carbon-intensive societies at a cost to our own health and that of our planet. However uncomfortable, interrogating and then dismantling those systems is precisely what is required to mitigate climate change the best we can, and adapt to the change that has already been 'locked in' by our inaction.

Both those involved in the health sector and those passionate people outside of it have a crucial role to play in ensuring healthy climate action for Aotearoa — changing systems from the inside out, as well as from the outside in. In this section, we'll delve into how we move from understanding to action; how the lens we use to understand climate change drives the actions we take and how effective they will be; and, finally, we'll introduce a few ways you can get involved in healthy climate action, with some key considerations for our health sector, government and communities.

THINKING CLIMATE JUSTICE

Taking meaningful action on climate requires first and foremost a recognition that our current predicament has been created, and enabled, to reach this critical point because of specific actions and inactions of governments, businesses, organisations and other decision-makers. While it is important on some level to take individual climate action, the reality is that some people have more power than others to influence this change. It is no surprise that those who have the most power, and who are responsible for the most climate-change-inducing actions, are also those who have been traditionally furthest from the frontlines and most removed from the realities our changing climate has on communities. In short, if we want to mitigate the worst effects of climate change, while also ensuring that future generations growing up in Aotearoa are healthier than our own, the bottom line is that we must take a *climate justice* approach, and remedy the injustices and imbalances in our society that have given rise to this problem in the first place.

When we talk about climate justice, we are talking about an approach and understanding of climate change that recognises climate change as a social and moral issue. We understand that climate change is the defining issue of our time, not because atmospheric carbon levels are inherently good or bad but because they ultimately affect people's lives and livelihoods. When we talk about climate justice we recognise that climate change has already been affecting many communities around the world — particularly Indigenous Peoples, those in less-wealthy countries, and other marginalised populations like people with disabilities, migrants and refugees, and sexual orientation and gender minorities. These communities are both disproportionately affected by climate change and also the experts on it, holding an immense wealth of knowledge on both the ways that climate change manifests in their regions and lives, and the systems and structures of society that produce climate change in the first place.

Climate justice as a concept and approach has evolved from

these communities, drawing on diverse knowledges and experiences to understand climate change in a human context. When we understand climate change in a human context, rather than an abstractly scientific one, we recognise not only the social structures and systems that underpin climate change — like capitalist imperatives towards extractivist industries and economic growth at all costs — but we can also recognise that as powerful as these systems are, they are fundamentally made up of human beings. Capitalism, colonialism and extractivism, as daunting and immovable as they may seem, are made up of human beliefs, norms, values and behaviours — and the good news is that these are entirely changeable.

Indigenous climate and health leadership

Western medical science and public health understandings are increasingly recognising and affirming the fundamental relationship between human health and the environmental and ecological contexts that we live in, and this is nowhere more evident than in ongoing research and advocacy on the interconnectedness between climate change and health. From individual health practitioners through to large United Nations bodies, the understanding of capitalism and colonialism as key root causes and drivers of climate change and ill-health are also being recognised.

Increasingly, this research and advocacy reinforces what many Indigenous communities, including tangata whenua in Aotearoa, have known for centuries. Many Indigenous communities, both globally and within Aotearoa, have significant bodies of knowledge relating to the environment, climate and health. While this may not be published in written books, research articles or western scientific volumes, it is often carried through stories, art forms, cultural practices, intergenerational teachings, and contemporary practice such as hunting or food gathering practices.

However, very often, this knowledge and expertise has been downplayed or regarded as 'unscientific' by western scientists, environmentalists and governments. As a result, this crucial knowledge has either been relegated to irrelevant mythology of 'uncivilised' peoples and been ignored completely, or, at the other extreme, has been approached as a knowledge resource to be extracted from without full, prior and informed consent of Indigenous Peoples, for the benefit of western academic portfolios. The devaluation of Indigenous and traditional knowledge is a function of racist colonialism and capitalism, which has prioritised western ways of knowing for the maintenance of the individualistic and extractivist systems that western development has benefited from.

Ultimately, these are the same systems that have driven the fossil-fuel-dependent and carbon-intensive development that is the underlying driver of climate change. They are also the same systems that have underpinned widening inequities in health, often prioritising individualised care for those who can afford it rather than more-widespread public health interventions and design that improves determinants of health for the whole population.

Unsurprisingly, given their colonialist and racist origins, these systems are widely evidenced as producing inequity particularly for Indigenous Peoples, people of colour, people with disabilities, refugees and migrants, and those with diverse sexual orientations, gender identities and sexual characteristics. At its core, ongoing colonialism and capitalism harms health, harms our climate, and disproportionally harms already marginalised and oppressed communities.

Indigenous climate knowledges

Acknowledging the diversity of Indigenous Peoples worldwide, and the fact that through processes of capitalism and colonisation many Indigenous people no longer reside in their ancestral homelands, Indigenous communities are typically best placed to provide insights into changes within local environments and ecosystems, which are markers of our changing climate. Information about ecosystem changes are able to be observed when communities have continuous or regular occupation of their lands, in a way that is rarely quantified or examined by western science. For example, Indigenous communities do not require a 10-year scientific study to tell them that certain species of fish are being affected by climate change, when they can observe over years, or even over generations, the changing distribution of those species, as they have to adapt their hunting practices. Within te ao Māori, many of our pūrākau, or stories, speak of migratory patterns for key bird species, or the conditions in which to grow or cultivate certain foods. These pūrākau disseminate important information about hunting and gathering kai across the generations, but can also be revisited to understand how these patterns may have shifted due to climate change.

Indigenous traditional knowledge and Indigenous worldviews tend to conceptualise the world in more holistic terms than western understandings, often prioritising relationships — both between humans and with the world around them. Our worldviews often see systems as interconnected, and hold space for complexity, which is essential in understanding the multifactorial drivers of climate change and developing robust action to prevent it. Similarly with health, Indigenous approaches often focus on the whole person, and the whole person in the context of their families and wider communities, rather than focusing on one body system or set of symptoms. When it comes to both climate and health, surface-level understandings give rise to surface-level solutions. Recognising the complex relationships between systems and drivers of ill-health and climate change is essential to improve both.

Colonisation has produced profoundly unjust circumstances for Indigenous Peoples around the world, and Aotearoa is no different. The ongoing process of colonisation in New Zealand has produced a deeply unequal society, with deeply unequal health outcomes. Māori are disproportionately represented in almost every negative health statistic, and are likely to have reduced access to determinants of health, including access to health services themselves. Many within our communities are still navigating intergenerational trauma that they have inherited through colonisation, as well as the contemporary racism it reproduces.

Colonisation at the root of climate and health inequity

Alongside the health harms outlined above, colonisation is also an extremely extractive and carbon-intensive process. Rather than embracing the diverse landscape and ecosystem of Aotearoa, settlers sought to re-create in many ways the systems of Europe (the failings of which were themselves drivers of this migration), importing plant and animal species that have had negative impacts on native flora and fauna, or requiring labour- and resource-intensive processes to maintain such systems — as evidenced by the incredibly intensive dairy industry which is a major contributor to New Zealand's emissions profile. The capitalist economic model that was similarly imported to Aotearoa also incentivises and prioritises extremely individualistic behaviours such as excessive consumption and hoarding of wealth and resources, which is incompatible with sustainable living and incompatible with the boundaries of our ecosystems and climate. The colonial society we have set up is simply not fit for this land, and it's not fit for our people either.

The colonial society we have set up is simply not fit for this land, and it's not fit for our people either.

Conversely, we see that Indigenous managed lands tend to have higher biodiversity, and Indigenous ways of living are much better suited to their unique landscapes and environments. For many Indigenous Peoples, the natural world is not regarded as something to be extracted from but instead something to be in relationship with, and we see this reflected in many cultural practices around the world, including tikanga Māori. For example, tangata whenua often view ourselves as kaitiaki, or stewards and protectors of our environment, taking only what we need when harvesting food, and being conscious to leave both breeding animals and juveniles to continue their life cycle and support sustainable populations. Principles like manaakitanga prioritise the sharing of resources, collective care, co-living, and other behaviours that also reduce our consumption and impacts on our environment and climate.

Learning from Indigenous climate action

While we don't wish to romanticise Indigenous Peoples and practices, and acknowledge that there were of course political and economic struggles within our pre-colonial societies as well, we can learn a lot from Indigenous worldviews and approaches to tackling the climate crisis. Considering the lessons and successes of many Indigenous societies in living vastly more sustainably than in our current western model, we can find comfort that dismantling the systems of oppression that drive climate change, poor health and a host of other harms is not only necessary work, but also work that can be profoundly meaningful and joyful. Rather than fearing the unknown, these approaches and ways of being, when led appropriately by Indigenous Peoples exercising their sovereignty on their lands and oceans, provide us with a hopeful and safe alternative to western capitalism and consumption.

Indigenous Peoples around the world and Māori here in Aotearoa have already been leading climate action for generations — protesting and preventing the construction of fossil-fuel pipelines, transforming our food systems through kai sovereignty and sharing of resources, reforestation and re-wilding of key ecosystems that function as 'carbon sinks', and reclaiming land and building sustainable co-living settlements. There are too many examples to list here, but many of these interventions not only help to prevent climate change but also provide for many of the determinants of health. Indigenous leadership on climate change provides a model for the kinds of climate action we can all be taking, and demonstrates that devolving power (including giving #LandBack) to Indigenous communities is a fundamentally necessary climate strategy both at local community levels and at the level of national government.

CLIMATE ACTION WITHIN THE HEALTH SECTOR

Health professionals and other members of our community who make up our health sector, from DHB board members to the Ministry of Health, administrative staff to frontline support workers, and nurses to

Māori health and allied health workers, all have major opportunities to contribute to healthy climate action. Health workers are often held in high regard by the general public and the communities they serve, and are therefore in a unique position to use their platforms and privileges to educate people on climate and health impacts, as well as directly advocating for change.

OraTaiao: The New Zealand Climate and Health Council is a not-for-profit organisation of over 600 health professionals and health professionals in training around Aotearoa, who work directly on the intersection of climate change and health. Members include nurses, occupational therapists, public health physicians, oncologists, surgeons, and many more doctors and allied health professionals who recognise the importance of climate change as both a challenge to our health and health system, and a crucial opportunity to build a healthier and more equitable society.

With a vision for 'Healthy Climate, Healthy People', OraTaiao has been instrumental in advancing research on climate and health impacts, and raising the consciousness of the health sector on this critical issue. Alongside other groups like the Sustainable Health Sector National Network, medical colleges and other professional networks, OraTaiao members have made submissions on a plethora of climate policies, influencing government at local and national levels to prioritise equity and health in climate action, with a focus on giving effect to Te Tiriti o Waitangi. This action includes advocacy for safe and accessible active and public transport, reduced agricultural emissions and a shift to more-plant-based diets, as well as successfully lobbying for the divestment of medical insurance companies from fossil fuels. OraTaiao has also actively campaigned for transparency in international agreements, like the Trans-Pacific Partnership Agreement and the Comprehensive and Progressive Agreement for Trans-Pacific Partnership, particularly highlighting the risk of climate-harming corporations blocking important public health and climate action through these international mechanisms. Many members of these climate and health organisations have also been instrumental in interrogating climate-harming practices within the health sector itself, such as hospitals' use of coal boilers and generators (OraTaiao, 2020).

There is an immense appetite for change within our health sector, in recognition of the inequity of current health outcomes as well as the risks associated with climate change. Organisations like OraTaiao regularly welcome new members from across the health sector and related fields to join campaigns and actions, and can also be supported to continue this critical work through their websites and social media platforms.

How can I take action?

We can take healthy and justice-focused climate action in many different ways — by divesting from extractive industries that harm our planet and health, disengaging from capitalism and consumerism, and dismantling systems of colonialism and white supremacy which keep Indigenous Peoples from exercising sovereignty and kaitiakitanga on their lands. Like climate change, climate action is a matter of degrees — every action we take to dismantle these systems is a reduction in our overall risk and the severity of climate and health impacts.

On a practical level, this action could look like:

- Switching to banks or other service providers that have policies against investing in or otherwise supporting fossil fuels.
- Reducing consumption from extractive industries by replacing car trips with active transport, or beef and dairy products with plant-based foods, if these are accessible and appropriate for you.
- Using land and resources that you have access to or control over for collective care — like setting up māra kai or food gardens in your front yard, and making food available through a pātaka kai or community pantry.

- If you own property, investigating how you can best use it to serve and provide for your community, including donating profits from investment properties to mana whenua or Māori organisations, or bequeathing the property to mana whenua.
- Using your professional power and resources to dismantle racism and excessive consumption within your own industry, including paying reparations or making donations to mana whenua or Māori organisations.
- Donating your time, funds or resources to support Indigenous sovereignty movements, or climate and health justice organisations.
- Collectively organising within your community to reduce individual consumption and develop health-giving relationships — for example sharing equipment with neighbours, having community meals, and cooking for or providing other supports to community members who are struggling.
- Writing submissions, lobbying government bodies and otherwise using existing mechanisms of local and central government to push for ambitious climate action, and to ensure that all policies take into account health, equity and rights enshrined in Te Tiriti o Waitangi.
- Educating yourself on white supremacy, colonialism, racism, capitalism, ableism, cissexism and other systems of oppression by engaging with books, articles, videos or other learning resources produced by marginalised authors and educators, and then acting to dismantle these systems within your own relationships and spheres of influence.
- Voting for candidates in community, local and national elections who prioritise Te Tiriti o Waitangi and equity, and those who support ambitious climate policies.
- If you are in a position of power — as a professional leader, business owner or someone with a public profile — using that position to cede power to those who are currently excluded

or under-represented. For example, passing on speaking or interview opportunities to colleagues from marginalised backgrounds; referencing and giving credit to Indigenous and other minority groups' work; hiring people from diverse backgrounds and doing the groundwork to ensure that workplaces are culturally safe and welcoming for them.

There are many ways to take action, and there will be a way to take action that is appropriate for you and your own situation, which speaks to your unique position, power and passions. You can think far beyond the scope of the list above and develop your own actions by considering a few key questions.

- What powers and privileges do I hold that can be leveraged in service of climate justice?
- Does this action address the root causes of climate change, rather than (or as well as) the surface manifestations of it?
- And lastly, is this action appropriate for me to take or lead given my position and privilege, and if not, can I instead support existing initiatives that are led by communities who are most affected?

Most importantly, make it a habit to regularly assess your actions and activism with a curious and open mind, and ideally with input from others around you. Be prepared to be challenged. Systemic change is not easy, so you need to be ready to be uncomfortable or give up power you've become accustomed to. Finally, make sure you are transparent and accountable in your activism, being open to change if you're made aware of problems in your approach, especially by marginalised groups. We will all make mistakes, and taking effective action on climate and health justice involves a lot of learning, un-learning and re-learning. The important part is that we commit ourselves to the process because we know that our goal is more important than our own ego or privilege.

Conclusion

In this chapter we have demonstrated that climate change has major impacts on human health, through a variety of mechanisms across direct, biologically mediated and social levels. We have also recognised that effective and meaningful action on climate change is not only critical to avoid devastating health impacts in the future, but can also offer many health gains in the here and now for us as individuals, whānau, communities and a broader society.

In coming to these understandings of the intersections between climate change and health, with a particular attention to health equity, we recognise that the root cause of climate change, and the root causes of poor health, are the same — mediated through systems of colonialism, extractivism, capitalism, ableism, classism and other oppressions. However, these systems are ultimately systems of human belief and behaviour, which can be changed through our collective action. Addressing the intertwined issues of climate change and health inequity at these roots is the most equitable way of making change, and is also far more effective, sustainable and cost-effective in the long term than simply addressing surface-level climate concerns. Creating a greener version of the status quo (like making individual passenger vehicles electric instead of fossil-fuelled) might help with our climate goals and targets in the short term, but it doesn't do anything to solve all the underlying systemic problems, and may actually increase inequity within our society.

Systemic change (like a massive transition to accessible active and public transport) helps to solve not only the emissions problem but all other problems associated with a car-dependent transport system. Likewise with our kai — wresting power from parasitic and extractive multinational corporations to ensure community control of food systems is not only good for climate, it's also good for almost every other metric of societal wellbeing. Whether we consider ourselves climate activists or concerned citizens and community members, we need to shift from individual action and surface-level interventions to systemic solutions. This may seem daunting, but is an exciting opportunity to not only prevent climate change but also create a fairer, healthier, more just society that we can all enjoy.

We have also discussed how our approach and understanding of climate change informs our action, and we have recognised that the tools and strategies for systemic change already exist within frontline communities. As the adage goes, those closest to the problem are also the closest to the solution. Centring Indigenous people and frontline communities, and devolving power to those communities, must be a key strategy at all levels if we are to tackle the climate crisis with any urgency and efficacy. This must occur in central and local government, within the health system, within the climate movement, and within our society in Aotearoa at large as a top priority.

As a nation, we have already been slow off the mark, relying on our small population, geographic isolation and the work of frontline communities to insulate us from climate change and its most serious effects thus far. We have missed the window for small, incremental and surface-level changes, and must commit to ambitious systemic change — and we must start now. The harms from climate change, especially on our health, occur as a matter of degrees. Every action we take in the right direction is a potentially life-saving vote for a healthy climate and a healthy society.

Bibliography

S. Al Mahri & A. Bouchama. Heatstroke in A. A. Romanovsky (ed). *Handbook of Clinical Neurology*, 2018, vol. 157, ch. 32, pp. 532–545. https://doi.org/10.1016/B978-0-444-64074-1.00032-X

Auckland Regional Public Health Service see ARPHS

ARPHS, 2020, *Salmonella*, <https://www.arphs.health.nz/public-health-topics/disease-and-illness/salmonella/>, 26 November 2020

R. Chapman, P. Howden-Chapman, H. Viggers, D. O'Dea & M. Kennedy. Retrofitting houses with insulation: a cost–benefit analysis of a randomised community trial. *Journal of Epidemiology & Community Health*, 2009, vol. 63, no. 4, pp. 271–277. doi: 10.1136/jech.2007.070037

C. A. Celis-Morales, D. M. Lyall, P. Welsh, J. Anderson, L. Steell, Y. Guo, R. Maldonado, D. F. Mackay, J. P. Pell, N. Sattar, J. M. R. Gill. Association between active commuting and incident cardiovascular disease, cancer, and mortality: prospective cohort study. BMJ, 2017, vol. 357, j1456, doi: https://doi.org/10.1136/bmj.j1456

A. Costello, M. Abbas, A. Allen, S. Ball, S. Bell, R. Bellamy et al. Managing the health effects of climate change. *Lancet*, 2009, vol. 373, no. 9676, pp. 1693–1733. https://doi.org/10.1016/S0140-6736(09)60935-1

Dahlgren G, Whitehead M. 1991. Policies and Strategies to Promote Social Equity in Health. Stockholm, Sweden: Institute for Futures Studies.

J. Drew, C. Cleghorn, A. Macmillan & A. Mizdrak. Healthy and climate-friendly eating patterns in the New Zealand context. *Environmental Health Perspectives*, 2020, vol. 128, no. 1, 017007. https://doi.org/10.1289/EHP5996

L. Feinstein, R. Sabates, T. M. Anderson, A. Sorhaindo & C. Hammond. What are the effects of education on health? In OECD, *Measuring the effects of Education on Health and Civic Engagement: Proceedings of the Copenhagen Symposium*. Copenhagen: OECD, 2006.

J. Franks. Raise benefits by Christmas, charities urge government, *Stuff News*, 9 November 2020. https://www.stuff.co.nz/national/123335339/raise-benefits-by-christmas-charities-urge-government

Georgetown University Medical Centre: Centre for Global Health Science and Security, 2018, *Establishing Infectious Disease Research Priorities In Areas Affected by Conflict*, <https://georgetown.app.box.com/s/qcye1xyifc2t487y1h4mp2akklgxldsz>, accessed November 21st 2020

R. Hond, M. Ratima & W. Edwards. *Global Health Promotion*, 2019, vol. 26, no. 3, pp. 44–53. https://doi.org/10.1177%2F1757975919831603

P. Howden-Chapman, J. Crane, A. Matheson, H. Viggers, M. Cunningham, T. Blakely et al. Retrofitting houses with insulation to reduce health inequalities: Aims and methods of a clustered, randomised community-based trial. Social Science and Medicine, 61 (2005): 2600-2610.)

HQSCNZ (Health Quality & Safety Commission). Asthma, 2020. https://www.hqsc.govt.nz/our-programmes/health-quality-evaluation/projects/atlas-of-healthcare-variation/asthma/#[REF], accessed 10 February 2021.

R. G. Jones. Climate change and Indigenous health promotion. *Global Health Promotion*, 2019, vol. 26, suppl. 3, pp. 73-81. doi: 10.1177/1757975919829713

R. G. Jones, B. Kidd, K. Wild & A. Woodward. Cycling amongst Māori: patterns, influences and opportunities. *New Zealand Geographer*, 2020, vol. 76, no.

3, pp. 182–193. doi: 10.1111/nzg.12280

M. K. Joy, K. J. Foote, P. McNie & M. Piria. Decline in New Zealand's freshwater fish fauna: effect of land use. *Marine & Freshwater Research*, vol. 70, no. 1, pp. 114–124. https://doi.org/10.1071/MF18028

J. P. Julian, K. M. de Beurs, B. Owsley, R. J. Davies-Colley & A.-G. E. Ausseil. River water quality changes in New Zealand over 26 years: response to land use intensity. *Hydrology and Earth System Sciences*, 2017, vol. 21, pp. 1149–1171. https://doi.org/10.5194/hess-21-1149-2017

C. H. Katelaris & P. J. Beggs. Climate change: allergens and allergic diseases. *Internal Medicine Journal*, 2018, vol. 48, no. 2, pp. 129–134.

N. Khetarpal & I. Khanna. Dengue fever: causes, complications, and vaccine strategies. *Journal of Immunology Research*, 2016, vol. 2016, article ID 6803098. https://doi.org/10.1155/2016/6803098

I. R. Lake, N. R. Jones, M. Agnew, C. H. Goodess, F. Giorgi, L. Hamaoui-Laguel, M. A. Semenov, F. Solmon, J. Storkey, R. Vautard & M. M. Epstein. Climate change and future pollen allergy in Europe. *Environmental Health Perspectives*, 2017, vol. 125, pp. 385–391.

G. Lindsay, A. Macmillan & A. Woodward. Moving urban trips from cars to bicycles: impact on health and emissions. *Australian and New Zealand Journal of Public Health*, 2011, vol. 35, no.1, pp. 54–60.

L. Luzatto. Sickle cell anaemia and malaria. *Mediterranean Journal of Hematology and Infectious Diseases*, 2012, vol. 4, no. 1, e2012065. doi: 10.4084/mjhid.2012.065

S. Mandic, A. Jackson, J. Lieswyn, J. S. Mindell, E. García Bengoechea, J. C, Spence, B. Wooliscroft, C. Wade-Brown, K. Coppell, and E. Hinckson. *Turning the Tide - from Cars to Active Transport*. Dunedin, New Zealand: University of Otago, 2019, retrieved from https://www.otago.ac.nz/active-living/otago710135.pdf

MFE, 2020, *New Zealand's Greenhouse Gas Inventory 1990-2018*, https://www.mfe.govt.nz/publications/climate-change/new-zealands-greenhouse-gas-inventory-1990-2018, April 2020.

T. Moeke-Pickering, M. Heitia, S. Heitia, R. Karapu & S. Cote-Meek. Food security and food sovereignty issues in Whakatāne. *MAI Journal*, 2015, vol. 4, no. 1. http://www.journal.mai.ac.nz/sites/default/files/MAIJrnl_V4Iss1_Pickering.pdf

MOH, 2020, *Giardia*, <https://www.health.govt.nz/your-health/conditions-and-treatments/diseases-and-illnesses/food-and-water-borne-diseases/giardia>, 12 August 2020

MOH (Ministry of Health). Health loss in New Zealand: a report from the New Zealand Burden of Diseases, Injuries and Risk Factors Study, 2006–2016. Wellington: MOH, 2013. https://www.moh.govt.nz/notebook/nbbooks.nsf/0/F85C39E4495B9684CC257BD3006F6299/$file/

health-loss-in-new-zealand-final.pdf

MOH (Ministry of Health). Health and Independence Report 2017. The Director-General of Health's annual report on the state of public health. Wellington: MOH, 2018. https://www.health.govt.nz/system/files/documents/publications/health-and-independence-report-2017-v2.pdf

New Zealand Human Rights Commission & He Kainga Oranga/Housing and Health Research Programme, University of Otago, Wellington. Inadequate housing in New Zealand and its impact on children: thematic snapshot report to the United Nations Committee on the Rights of the Child. Wellington: HRC, 2016 https://www.hrc.co.nz/files/7014/7407/6639/Thematic_snapshot_report_of_NZHRC_for_UNCRC_73rd_session_final.pdf

N. H. Ogden, J. K. Koffi, Y. Pelcat & L. R. Lindsay. Environmental risk from Lyme disease in central and eastern Canada: a summary of recent surveillance information. *Canada Communicable Disease Report*, 2014, vol. 40, no. 5, pp. 74–82.

OraTaio. 2020. Ora Taiao, Phasing out coal in schools and hospitals a win-win for health and the climate. *Ora Taiao*, 30 January 2020. https://www.orataiao.org.nz/phasing_out_coal_in_schools_and_hospitals_a_win_win_for_health_and_the_climate, accessed 21st November 2020.

B. J. Phiri, A. B. Pita, D. T. S. Hayman, P. J. Biggs, M. T. Davis, A. Fayaz, A. D. Canning, N. P. French & R. G. Death. Does land use affect pathogen presence in New Zealand drinking water supplies? *Water Research*, 2020, vol. 185, 116229. https://doi.org/10.1016/j.watres.2020.116229

C. Shaw, E. Randal, M. Keall & A. Woodward. Health consequences of transport patterns in New Zealand's largest cities. *New Zealand Medical Journal*, 2018, vol. 131, no. 1472, pp. 64–72.

K. R. Smith, A. Woodward, D. Campbell-Lendrum, D. D. Chadee, Y. Honda, Q. Liu, J. M. Olwoch, B. Revich & R. Sauerborn. Human health: impacts, adaptation, and co-benefits. In C. B. Field, V. R. Barros, D. J. Dokken, K. J. Mach, M. D. Mastrandrea, T. E. Bilir, M. Chatterjee, K. L. Ebi, Y. O. Estrada, R. C. Genova, B. Girma, E. S. Kissel, A. N. Levy, S. MacCracken, P. R. Mastrandrea & L. L. White (Eds), *Climate Change 2014: Impacts, Adaptation, and Vulnerability. Part A: Global and Sectoral Aspects. Contribution of Working Group II to the Fifth Assessment Report of the Intergovernmental Panel on Climate Change*, pp. 709–754. Cambridge: Cambridge University Press, 2014.

M. Springmann, K. Wiebe, D. Mason-D'Croz, T. B. Sulser, M. Rayner & P. Scarborough. Health and nutritional aspects of sustainable diet strategies and their association with environmental impacts: a global modelling analysis with country-level detail. *Lancet Planetary Health*, 2018, vol. 2, no. 10, pp. E451–E461. https://doi.org/10.1016/S2542-5196(18)30206-7

Statistics New Zealand, *New Zealand's greenhouse gas emissions*, https://www.stats.govt.nz/indicators/new-zealands-greenhouse-gas-emissions, accessed 25 January 2021

B. A. Swinburn, V. I. Kraak, S. Allender, V. J. Atkins, P. I. Baker, J. R. Bogard . et al. The global syndemic of obesity, undernutrition, and climate change: The Lancet Commission report. *The Lancet Commissions*, 2019, vol. 393, no. 10173, pp. 791–846. https://doi.org/10.1016/S0140-6736(18)32822-8

D. Tilman & M. Clark. Global diets link environmental sustainability and human health. *Nature*, 2014, vol. 515, issue 7528, pp. 518–522. doi: 10.1038/nature13959

United Nations Sustainable Development Goals, 4. Quality Education, https://www.un.org/sustainabledevelopment/education/, accessed 21 November 2020

A. M. Vicedo-Cabrera, M. S. Ragettli, C. Schindler & M. Röösli. Excess mortality during the warm summer of 2015 in Switzerland. *Swiss Medicine Weekly*, 2016, vol. 146, w14379. https://doi.org/10.4414/smw.2016.14379

C. M. Walter, E. K. Schneider-Futschik, L. D. Knibbs & L. B. Irving. Health impacts of bushfire smoke exposure in Australia. *Respirology*, 2020, vol. 25, pp. 495–501. doi: 10.1111/resp.13798

N. Watts, W. N. Adger, P. Agnolucci, J. Blackstock, P. Byass, W. Cai et al. Health and climate change: policy responses to protect public health. *Lancet*, 2015, vol. 386, no. 10006, pp. 1861–1914. https://doi.org/10.1016/S0140-6736(15)60854-6

W. Willett, J. Rockström, B. Loken, M. Springmann, T. Lang, S. Vermeulen et al. Food in the Anthropocene: the EAT–*Lancet* Commission on health diets from sustainable food systems. *The Lancet Commissions*, 2019, vol. 393, no. 10170, pp. 447–492. https://doi.org/10.1016/S0140-6736(18)31788-4

N. Wilson, D. Slaney, M. G. Baker, S. Hales & E. Britton. Climate change and infectious diseases in New Zealand: a brief review and tentative research agenda. *Reviews on Environmental Health*, 2011, vol. 26, no. 2, pp. 93–99. doi: 10.1515/reveh.2011.013

L. Wiltshire. Digital divide: Thousands of students now able to access internet from home. *Stuff News*, 11 August 2020. https://www.stuff.co.nz/national/education/122407001/digital-divide-thousands-of-students-now-able-to-access-internet-from-home, accessed 21st November 2020

World Forum for Food Sovereignty, 2007, 'Declaration of Nyéléni', *Declaration of the World Forum for Food Sovereignty*, Sélingué, Mali. retrieved https://nyeleni.org/spip.php?article290

Y.-F. Xing, Y.-H. Xu, M.-H. Shi & Y.-X. Lian. The impact of PM2.5 on the human respiratory system. *Journal of Thoracic Disease*, 2016, vol. 8, no. 1, pp. E69–E74.

Nothing about us without us: Climate change and disability justice

JASON BOBERG
SustainedAbility Disability Climate Network
& Creative Director, Activate Agency

KERA SHERWOOD-O'REGAN
SustainedAbility Disability Climate Network
& Impact Director, Activate Agency

'Nothing about us without us' is a catch-cry of the disability rights movement. It is a slogan that has been shouted by our disabled predecessors around the world, on issues from eugenics to employment discrimination, mass institutionalisation, inaccessible roadways and obstructive urban design. Many of the freedoms and accessibilities enjoyed by wider society have been built thanks to disabled activists who have shouted this slogan and campaigned to be included in decisions that affect our lives, realities and bodies.

Climate change is about us. Through both direct and systemic means, disabled people are disproportionately affected by our changing environment, and this is magnified for those who also experience intersecting marginalisation across the lines of race, indigeneity, gender, sexual orientation and migrant status. Yet despite how critical the impact of climate change is on our communities, we are frequently left out of conversations and decision-making on this important issue. The implication of these discriminatory attitudes and assumptions is that we, as disabled people, are passive victims rather than active agents of change who have something to offer. This is patently, not to mention morally, wrong. However, when climate action includes diverse disabled people and our perspectives, we are able to innovate solutions that not only are appropriate for our communities but also provide solutions for our whole society. In this chapter we'll discuss some of the ways in which climate change affects our community, as well as disabled movements we can draw on for climate action, and some suggestions for how disabled and non-disabled readers can take action for a more inclusive, accessible and just climate movement.

While it's clear that our community is severely affected by climate change, and has been made vulnerable to its impacts through ableist social structures, policies and planning, disabled people also have a long legacy of problem-solving and change-making. We are experts at coming up with solutions to the issues that affect us, in many cases out of necessity having to design alternative systems, tools and other innovations to get access to what we need, and to get by in a world that is frequently hostile to our bodies and identities. Our community members are experts on all variety of specialities required to mitigate climate change, and ensure that climate adaptation is inclusive and accessible for everyone — from disabled policy analysts to politicians and diplomats, to climate experts, community organisers, campaigners and activists.

We cannot get out of the climate crisis with the same thinking, models and decision-makers that got us into it in the first place. The inclusion and leadership of disabled people in climate change decision-making and action is crucial to innovate solutions to this global problem that ensure the world we create is fairer and more equitable for everyone.

Disability rights and justice

Over 1 billion people — 15 per cent of the global population — experience some form of disability, making us the world's largest minority group (World Health Organization, 2011: 29). One in four New Zealanders is disabled (Statistics New Zealand, 2014: 2). Encompassing those with physical, intellectual, sensory and mental disabilities and impairments, as well as chronic illnesses, disability and disability rights affects a broad range of people and even includes those who may not identify with the term 'disabled' but who are nonetheless protected by disability rights for which the community has fought. Many more people still will find themselves disabled over the course of their lifespan due to injury, illness and advancing age.

Impairments that come within the scope of disability are conditions, illnesses or injuries that cause a difference or loss of function; however, it is vital to note that these are defined for legal purposes and are measured against a benchmark of a supposed 'normalcy' which in itself is problematic. Examples of impairment are vast and include limb differences, blindness or vision impairment, chronic illnesses, and mental health conditions. Disability is an evolving societal concept: there is no master list of what is or isn't counted, and this is important. Even the official definition under the United Nations Convention on the Rights of Persons with Disabilities (UNCRPD) is deliberately vague to avoid 'exclusion by inclusion', or a risk that only those impairments listed are considered 'disabled enough' to warrant protection under these rights. The definition states that.

> *Persons with disabilities include those who have long-*
> *term physical, mental, intellectual, or sensory impairments*
> *which in interaction with various barriers may hinder*
> *their full and effective participation in society on an*
> *equal basis with others (United Nations, 2006: 4).*

The UNCRPD further specifies the rights that disabled persons are entitled to, including equality and non-discrimination, accessibility — the ability 'to live independently and participate fully in all aspects of life' (United Nations, 2006 : 9) — and participation in political and public life. However, despite being both a signatory to the UNCRPD and considering ourselves a progressive and fair democracy, New Zealand still has major issues regarding upholding the rights of our community, including having a lower minimum wage for disabled workers than non-disabled workers; lack of accessible transport and housing; and a 'two-tiered' disability system where previously non-disabled people who acquire a disability due to accident or injury receive compensation through the Accident Compensation Corporation (ACC), while those who are born disabled or become disabled by sickness or life course receive comparatively less support through the Ministry of Health. Disabled people in New Zealand continue to experience discrimination across society, including when it comes to climate change and climate action.

DISABILITY IDENTITY AND TERMINOLOGY

This chapter refers frequently to the challenges that lie ahead for disabled people as a broad community. While there are many shared experiences and struggles that bring us together as disabled people, it is important to recognise that our community is not homogenous — we all have different experiences, perspectives and access requirements that need to be considered.

Therefore, when we are thinking and talking about emergency situations and other ways in which climate change affects our community, it is important to highlight that it is not only about physical disabilities and access — like, say, accessible ramps for emergency exits. It is crucial that we also include, in our planning, action and responses, people with psychosocial disorders and invisible disabilities, such as post-traumatic stress disorder (PTSD), autism or depression, as well as those who claim other labels for themselves but also experience discrimination through ableism and are entitled to protection under disability rights.

The authors acknowledge the diversity within our disabled communities and understand that choices around terminology and labels are often deeply personal, and connected to deep traditions within our communities, impairment groups and cultures. For this reason, we use the term 'non-disabled' in this chapter, rather than 'able-bodied', in order to acknowledge the rights and needs of all disabled people.

The term 'disabled person' is preferred by many disabled people, particularly those with strong connections to disability rights and justice movements, because it acknowledges disability as a core part of our identity. For many, the authors included, identifying as disabled is a profoundly liberating experience — connecting us with others around the world and providing us with a framework to understand that our bodies and minds are not problematic, but that it is the structures of an ableist and exclusive society that cause us harm.

We will also refer in some cases to 'people with disabilities' or PWD — terminology which comes from the social model of disability and the UNCRPD. This terminology is common in international legal spheres and health organisations, and is also the preferred terminology from some community and impairment-based groups. Some communities including many d/Deaf people prefer to identify simply as Deaf, highlighting the rich culture of their community and language, and may not use the term 'disabled' for themselves.

Euphemisms such as 'special needs' or 'differently-abled' are widely deemed to be offensive by many disabled people, and we would encourage avoidance of this language unless people identify this way themselves.

The authors identify proudly as disabled people ourselves, and use this term with intention to de-stigmatise and reclaim the word 'disability', and to highlight the whakapapa of the knowledges we are privileged to share in this chapter thanks to the support, encouragement and education from our beautiful and strong community of disabled activists, scholars and change-makers in Aotearoa and around the globe.

However, we recognise that, due to ableist societal pressures and judgement, for some people it's simply not safe or helpful to identify as disabled or to disclose their impairments, and for others there may simply be other labels or identities that they prefer. Therefore, while we use 'disabled' and 'non-disabled' terminology in our own writing, we maintain that it is the right of people themselves to claim the label and identify in the way that is best for them — and as a general rule, non-disabled people should mirror the terminology used by that person, or ask where appropriate how they would like to be identified.

DISABILITY MODELS AND MOVEMENTS

To understand how disability rights are crucial to people in New Zealand, and how climate policies and action can include us, first we must understand the evolution of the conceptualisation of disability. There are two prevailing and starkly different models of disability: medical and social.

Today, the medical model of disability is widely considered offensive and outdated by disabled communities and the disability rights movement globally. Nevertheless, it is important to understand this model to recognise how disabled people's realities have been shaped through successive governments and policies, and how this historical model leads to the kinds of discrimination that leave disabled people so disproportionately affected by climate change. The medical model is concerned with impairments or differences in the function of disabled people's bodies compared with a non-disabled norm. This model sees that a person's disability is inherently an undesirable defect and problem, and has deep roots in industrialisation and our current capitalist economic model. When disabled people couldn't participate on an equal level to non-disabled workers in factories and other economic endeavours during the Industrial Revolution, they were often seen as expendable or burdens to society. Under a medical view, this supposed burden required medical intervention to 'fix' people's disability so they could participate economically, or was alternatively used as a reason to exclude disabled people from society. This model of understanding disability is inherently ableist and harmful, and has been a primary driver of eugenics movements and policies that forced sterilisation of disabled people to 'breed out' disability, as well as mass abuse and exclusion of disabled people from society through institutionalisation — practices now widely recognised as a grave abuse of human rights but which still have a painful and harmful legacy on our community today.

The social model of disability was developed in efforts to move away from the medical model and is the current working concept for many organisations, including the United Nations, disabled people's organisations, and some of the not-for-profit sector. Although it is considered best practice in these spheres, the social model is still being adopted by the wider health sector and related organisations. Rather than identifying the body-minds of disabled people as a problem, the social model highlights systemic barriers, attitudes and social exclusion as the problems to be addressed. The social model posits that it's not the person who is inherently disabled, but society that is disabling for a person with an impairment.

For the purposes of this chapter, we won't go into depth about the history of our disability rights movements or into these models of disability. However, to help us understand disability rights and justice advocacy within the context of climate change and climate activism, it is important to briefly touch on the disability rights and disability justice movements.

A movement for disability rights

Building from the social model — which identified the environment and societal barriers as the problem, rather than disabled people ourselves — disability rights movements have emerged in many countries around the world. These were, and still are, broad civil rights movements that operate not only on a political level, challenging policies that breach our rights, but also through building a strong 'on the ground' ecosystem, with connections between grassroots organisations and activists, community networks and formally mandated disabled people's organisations and other bodies. This movement has been key in establishing a number of disability rights precedents globally, and also in progressing an understanding of disability as an identity, building a strong culture of disability pride that rejects deficits-based understandings of our communities, minds and bodies.

The Disability Rights movement has earned many of the rights and protections that are now essential to build upon for the protection of disabled people in the context of climate change. This includes being able to use mechanisms of the United Nations to put pressure on governments who violate disability rights or who do not respect our right to be included in climate mitigation and adaptation planning; and also provides us with tools to advocate at a local community level, where local and national governments have signed on to responsibilities and standards towards disabled people's rights and engagement, and can therefore be required to consult with our communities and consider our rights and needs in any local climate policy or policy which may have an implication on our rights and experience of climate change.

FROM RIGHTS TO JUSTICE

While the Disability Rights movement has provided our community with many tools for advocacy and action, including on climate change, it is important to recognise that it only goes so far. Key critiques of the movement include that it has traditionally been, and still is to this day, dominated by more-privileged voices and perspectives within the disabled community, resulting in policies and campaign 'wins' that disproportionately serve White, cis-gendered heterosexual men with disabilities, particularly those with economic, class and academic

The Disability Rights movement has not always made space for the perspectives of disabled people who also experience intersecting marginalised identities.

privilege. Many of the mechanisms of disability rights and advocacy are therefore particularly elitist, and difficult for those who are not academically, financially or otherwise privileged to access for their own and their community's advocacy and justice.

The Disability Rights movement has not always made space for the perspectives of disabled people who also experience intersecting marginalised identities, such as those who are also Indigenous, Black, People of Colour, Queer, transgender, migrants or refugees, and those who are fat or older. The result has been that those members of our communities who are most marginalised, and therefore most at risk from a variety of socioeconomic issues and the most severe impacts of climate change, have often had their perspectives and needs ignored or sidelined in favour of claiming progress for the most privileged members of our community.

Many multiply-marginalised disabled activists and change-makers have therefore moved towards a movement of 'Disability Justice', with a framework developed by the disability group Sins Invalid that explicitly centres the voices, experiences, needs and liberation of multiply-marginalised disabled folk. Disability Justice also recognises the whakapapa or genealogy of other intersecting movements for rights and liberation, including those movements for Indigenous sovereignty, Queer liberation and anti-racism, providing a more diverse and interconnected platform for change.

Applying a Disability Justice lens to our climate action enables us to recognise that the root causes and drivers of climate change are also the drivers of other systems of oppression, including racism, colonialism, cissexism and classism. Taking a Disability Justice approach to climate change and climate action requires us to approach our advocacy, campaigning, policy and direct action with multiply-marginalised people in mind, and implores us to take action on these root causes — dismantling systems of oppression entirely, rather than merely making oppressive systems less carbon-intensive. Applying a Disability Justice lens to our climate action reminds us that there are many experiences of disability, and that we must design our actions and responses with broad input and broad experiences in mind rather than looking for a quick 'disability check box' to tick. It requires us to recognise how systems of oppression intersect and uphold each other, each further locking us into lifestyles, ways of being and modes of thinking that drive climate harm, and harm to our communities. It is the opinion of the authors that a Disability Justice approach is a fundamentally necessary prerequisite for ethical, ambitious and meaningful climate action.

How climate change affects disabled communities

CLIMATE DISASTERS

Now that we've established some of the working models and ways of understanding disability, we'll examine why these rights are so important when it comes to climate change. When fires, floods and other natural disasters occur, disabled people are among the worst affected. We are frequently left out of disaster planning, and so are often trapped by inaccessible escape routes and environments planned with only non-disabled people in mind. When it comes to rescue operations and allocation of resources in a civil defence emergency, disabled people are frequently assigned a low triage priority, meaning that many more suffer the most severe repercussions on our health, and on our social and economic situations. This is highlighted in the outcomes of the November 2018 Camp Fire in California, where the average age of the 85 victims was in the seventies and many were disabled (Tucker et al.; Wong, 2018). Factors often considered in triage situations, even unofficially, include a patient's life expectancy and the degree of disability they are expected to have after a medical intervention (Roth et al., 2018). Whichever way you turn it, and whether this is formalised in official health and emergency plans or left to the judgement of individual health professionals, the assumption is that a disabled life is of inherently poorer quality and therefore less worth saving. This reflects societal attitudes and structures that continually devalue disabled people and our experiences, reducing us to a one-dimensional trope of 'poor disabled person' without considering the many positive experiences and the value our lives have, and even the ways that experiencing disability can be beneficial in our lives. The Camp Fire represents ableism at both a structural and an individual level, where disaster response training reinforces triage hierarchies that put disabled people at greatest risk. If we continue to take a 'business as usual' approach to determining which lives are worth saving in climate disasters, these disabled deaths are the canaries in the coal mine.

The phenomenon of leaving disabled people out of disaster responses, and literally leaving us behind to die, is so commonplace even in so-called developed societies that the 'Right to be Rescued' has not only been coined as a popular campaign slogan but has even been successfully established as a legal right within certain jurisdictions in the United States, through a number of class-action lawsuits brought in order to give specific protections to disabled people in disaster contexts (Weibgen, 2015: 2465). We saw the need for these protections close to home in the 2020 Australian bushfires, when disabled citizens were forced to decide whether to risk leaving their accessible homes, essential aids and equipment to be packed into an emergency shelter that was inaccessible and ill-prepared. Following the fires, the New South Wales government funded a three-year research programme working with the disability community to improve disaster planning and recovery inclusiveness (E. Young, 2020).

When floods and fires inevitably increase in New Zealand, how well prepared will we be? Today, while building codes include suggestions for accessible emergency exits, their working definition of accessibility doesn't go far enough. For example, for somebody who uses a power wheelchair, their fire escape plan from a multi-storey building might look like some sort of outdated trolley system which nobody is practised in using, or they may simply be forced to find a 'safe place' to wait for emergency services and hope that these services arrive in time.[1] There is also a difference between policy and practice, where in practice many disabled people are themselves considered 'hazards' (i.e. using crutches or a cane) and are told to wait for services. In an emergency fire drill of a public building, one of the authors of this chapter was advised to stay behind and wait to be rescued, as their use of crutches was considered a hazard and might slow other evacuees down. Public data on how many people are at risk in such situations is not available; there is a dearth of statistics on issues affecting people with disabilities, a problem that the UN is addressing with their Disability Statistics programme (UNStats, 2021). But as floods and fires remain a relative likelihood in a climate-affected future, we can look overseas as a warning and as a strong reason for our own government to fund research in escape and recovery planning for disabled people, especially in preparation for a rise in climate-related disasters.

---

1   Ministry of Business, Innovation and Employment. 'Means of escape', https://www.building. govt.nz/building-code-compliance/d-access/accessible-buildings/means-of-escape/, accessed 31 January 2021.

HEALTH IMPACTS

In addition to the major concern over disaster response, there are also health effects of climate change on disabled people to consider. These are discussed in the previous chapter but some examples include that temperature rises can disproportionately affect people with spinal cord injuries that impair sweat function and affect the body's ability to respond to and regulate heat. Increased pollen from changing climate systems can also exacerbate respiratory conditions (Harrington, 2019), along with the increases in pollutant or particular matter that we would expect to occur with current patterns of emissions, and in cases of fire. It is important to remember that such health impacts, as illustrated in the previous chapter, tend to disproportionately affect people with experiences of marginalisation — including people who are disabled. We must therefore anticipate that such health impacts will be magnified for our community who are already disabled, as well as being a cause of new disability.

CLIMATE CHANGE MAGNIFIES INJUSTICE

Disabled people's risk in the face of climate change has much less to do with the function of our bodies and our minds than it does with decades of systemic oppression that perpetuates societies and environments that are hostile to our community. As highlighted by Weibgen (2015: 2408), natural disasters and their impacts are not merely a function of climate and weather systems, but also of human systems and the choices we make about how our societies should be structured, where we should live, and more: 'In this way, disasters are socially constructed. How we choose to respond to the urgent human needs that arise from large-scale weather events determines the degree to which these events become "disasters".'

Accordingly, climate change presents a 'threat multiplier' to disabled communities, magnifying the impacts of existing injustices that have negatively affected disabled people for centuries. Generations of socioeconomic marginalisation means that disabled people are more likely to be in precarious living situations — people with disabilities are twice as likely to experience poverty as our non-disabled counterparts (National Council on Disability, 2017: 11). With lower incomes often comes poorer quality of housing, and living in accommodation that is protective against health impacts of climate change — such as well-ventilated homes that offer some protection against rising temperatures — may be a choice that only the most privileged members of our community can afford. This is not to mention the challenges presented by relocations following disasters. It is estimated that only 2 per cent of New Zealand's existing housing stock, both state and privately owned, is accessible (Bhatia, 2019); at the time of writing, there are 1050 Kiwis on the waiting list for accessible state homes, with a staggering lack of accessible new builds on the horizon (1News, 2020). As extreme weather becomes more common, disabled people are less able to uproot our housing and so could be 'forced to remain in degraded environments without housing, employment, support networks or healthcare services' (Schulte, 2020). At present there is nothing on the table from our New Zealand government to ensure that our country will meet the demand for accessible housing.

Access to power

In the event that flooding, fires or other disasters cause power outages, again we can look overseas to understand how these might affect disabled people and what we can do to prepare for similar scenarios here in Aotearoa. In 2018, the biggest electricity provider in the US state of California, Pacific Gas and Electric, shut off the electrical grid to San Francisco and parts of the wider Bay Area as — due to their negligence in adequately maintaining the energy grid — it presented a hazard to the rapidly spreading wildfires in California. The warning time between their decision and the shutdown was a matter of days, and the result of their criminal negligence was the deaths of 84 people in the fire and at least one confirmed death of a disabled person due to the shutdown itself. For users of aids like motorised wheelchairs and oxygen ventilators, the shutdown signified a matter of life or death (Ho, 2019). Late Disability Justice activist Stacey Park Milbern, alongside other proudly fat, queer, older, People of Colour and disabled people, formed the public campaign 'Power to Live' to raise funds to support mutual aid networks, source generators, supply medication and ice, check in with friends and those in the community and find housing where needed, to get their community through the shutdown in the absence of support from both Pacific Gas and Electric and the state government (Green, 2019). In New Zealand we do not have statistics on how many people rely on ventilators or other medical and accessibility devices that require electricity, but it is safe to say that few members of our community own generators or would have the financial resources to buy one in the event of such emergencies. Although not on the scale of the issues in the United States, power outages as a result of extreme weather in New Zealand similarly put disabled people at increased risk.

Health emergencies

As the patterns of viruses and disease is expected to change alongside changes in climate, we need only look at the 2020 Covid-19 pandemic response internationally and in Aotearoa to imagine the effects of such health emergencies on disabled people. The Covid-19 example is particularly important for two reasons. The first is that while the government claimed to have considered the effects of lockdown and the pandemic on the disabled community, in reality we had limited legal safeguards and no transparency around the kinds of decisions that we witnessed being made overseas and which threatened to reach New Zealand. In the United Kingdom, many in the disabled community reported receiving 'do not resuscitate' (DNR) letters from their doctors, asking them to pre-emptively sign away their right to be resuscitated in the event that they were 'competing' at hospitals with non-disabled people for intensive Covid-19 medical treatment — again highlighting the triage priorities ingrained in the current health system. While some could refuse to sign, this is a clear example of medical coercion, undermining disabled people's right to make independent choices and give informed consent for their medical treatment. Furthermore, almost one-fifth of members of Learning Disability England reported unlawful DNRs appearing in their medical records without their knowledge or consent (Lintern, 2020), and there is community knowledge of this happening to New Zealanders in Aotearoa. These events highlight how quickly disabled and high-risk people are deemed expendable in health emergencies, and is a major cause for concern given the changing distribution of communicable diseases likely with changing climate patterns, as well as other health emergencies likely to occur due to climate change.

The second reason why Covid-19 is so relevant to Disability Justice in a future with changing patterns of infectious disease is that the medical consequences of Covid-19 are increasing impairments and disabilities worldwide. 'Long Covid' is a term that is still being defined, but many Covid-19 survivors report long-term symptoms such as chronic fatigue and pain, respiratory issues, migraines and brain fog, which many have pointed out bear striking similarities to chronic fatigue syndrome, fibromyalgia and other recognised chronic disabilities. This is an evolving situation, but in the next World Health Organization statistics we will see how significantly the global disabled population will have increased due to Covid-19, and can similarly expect climate change to alter the demographics of our disabled population as people acquire disability due to injury in disaster events; respiratory conditions due to pollution and fire smoke; and mental distress from climate impacts socially, economically and, especially, culturally.

Unintended harm of climate action

As if the disproportionately harmful effects of climate change on disabled people are not already frightening, the present climate action being taken is, and will — without leadership, involvement and governance from diverse and multiply-marginalised disabled communities — further create barriers and remove access to the services and solutions we have fought hard for. There is a name for climate action that doesn't respect disabled people and our rights: eco-ableism. Simply put, this is when actions that are supposed to cut climate emissions or pollutants and other environmentally-harming behaviour also undermine disability rights. Without climate organisations, activists and policy-makers applying a Disability Justice lens or including disabled people in the conversation to uphold disability rights, there is a genuine risk of further entrenching the marginalisation of disabled people, magnifying inequities and even causing explicit harm, especially for disabled people who are also Indigenous, queer, people of colour, migrants or have other experiences of marginalisation.

CLIMATE CHANGE COMMISSION

In 2019 the New Zealand government introduced the Zero Carbon Act, which aims to legally bind the government of the day to creating long-term plans for a nationwide climate change mitigation strategy, which should in turn allow for better planning for climate change from government across all its decision-making. A major outcome of the Zero Carbon Act was the establishment of the Climate Change Commission. However, the Act provides no explicit mandate to engage with disability organisations or disabled people as key stakeholders in climate action. Despite advocacy from the disabled community, including SustainedAbility — our climate and disability network — it appears that our voices continue to go unheard. Under New Zealand's obligations as a signatory to the UNCRPD, it's clear that this Act and the Commission's terms of reference must be updated sooner rather than later to give explicit mandate to take into account disability rights and justice. To fully implement the UNCRPD, we must make sure that principles of Disability Rights are enshrined in law wherever relevant, before further climate-related policies are brought in that could actively harm our community.

TRANSPORT AND OUR BUILT ENVIRONMENT

One key policy area where eco-ableism often appears is transport. Reducing carbon emissions is great; reducing disabled people's access to participation in our society is not. For people with physical disabilities, public transport is very often inaccessible and highly inconvenient. So-called 'green solutions', like eco-oriented urban planning and public transport, continue to be designed without disabled people in mind. The result is that services are not suitable for, or accessible to, disabled people, our community is often criticised when we must then rely on carbon-intensive alternatives like driving, and in some cases access that already exists is even removed, such as the removal of car parking spaces that we might need to be sufficiently close to services.

This touches on something fundamental to disability rights history: the 'curb-cut effect'. In 1970s United States, when disabled people took action during the horror of institutionalisation, we fundamentally realised that cities had not been built for us. Just one of the many issues was raised curbs protruding from the street — prohibitive to people with disabilities, particularly wheelchair, cane and crutches users, and those with gait or other movement impairments. In the years that followed, disabled activists took hammers to the curbs and used bags of concrete to create what we now think of as something completely ordinary (if we think of it at all) — the 'curb-cut' slope from the footpath to the road, used every day by everyone from cyclists to people pushing prams or hauling suitcases. Just as the first touchscreens were designed for and by the disabled community but are now present on your mobile phone, innovations led by disabled people often make their way into the mainstream, very often also to the benefit of non-disabled people (Jackson, 2018). When we think of eco-friendly urban design, instead of planning pedestrianised spaces that exclude disabled people from participation, disabled people should be invited to sit at the table and do what we have done in the past to improve the situation for *everybody*: innovate. If we're serious about climate action, can we really afford to leave a quarter of the population locked into climate-intensive systems?

A key missed opportunity in climate-related action in Aotearoa relates to the new electric buses on the roads in Auckland, which are built to the same specification as existing petrol-powered buses (Ternouth, 2020) and replicate their seating, ramps and physical safety issues. Likewise, in Christchurch the council spent $94,000 designing a new colour and concept for their fleet of buses and, as a result of excluding disabled people from discussions, selected a colour that many visually impaired people struggle to see (C. Young, 2020). It is an absurd situation: cities and transport must be designed for one another, but neither is designed with disabled people in mind — leaving disabled people unable to access many city services, and locking us into more carbon intensive transport. As a move to electric vehicles is imminent, as part of the very necessary action to prevent climate change, it is critical that we seize the opportunity to ensure that new infrastructure and public services are accessible to disabled people without added costs and challenges to availability. There are still improvements required to make car-friendly infrastructure for people with disabilities, and any actual solution will be radically outside of what is currently under discussion in council chambers. In order for environmental action to include disabled people and create more-accessible transport for everyone, we need to offer choices — not just a single alternative — and disabled people are best placed to suggest accessible options.

INDIVIDUAL ACTION AND THE WAR ON PLASTICS

An additional example of eco-ableism in climate and wider environmental spheres is marked by the so-called 'plastics debate'. For decades, mainstream climate conversations have centred around individual lifestyle choices and sacrifices to reduce waste, pollution and other consumption behaviours like switching to more 'eco-friendly' light bulbs and taking short showers. No discussion has been more prominent than that of plastic pollution. For many disabled people, however, reducing plastic consumption at an individual level only creates new challenges and harms.

The New Zealand government's announcement to phase out the use of single-use plastic items like drinking straws (Flaws, 2020) is low-hanging fruit to show goodwill to environmentalists at the cost of an essential tool that some disabled people rely on to feed themselves.

> **Disabled people are already climate leaders; but the public is only just beginning to shift their perception of our community and create a space for us to participate meaningfully in demonstrations and consultation processes as our full, glorious disabled selves.**

Despite continuous advocacy by the disabled community, including clear case studies and demonstrations of why single-use plastic straws are the only accessible option for many people with motor disabilities who rely on the material's flexibility to eat and drink, governments around the world have proceeded with straw bans. Of course it is important to reduce global reliance on plastics, but it is estimated that plastic straws account for only 0.02 per cent of plastic debris in the ocean (Jambeck et al., 2015). Taking these away from disabled communities who need them to survive is hardly the best-value proposition for removing plastic pollution and reducing dependence on oil. Overseas examples of the straw ban indicate that it also forces disabled people to disclose their disability to, for example, social peers and wait-staff in order to get access to straws that are not freely available at eateries. This puts the access of disabled people to essential nutrition, and to their right to equal enjoyment of society, in the hands of restaurant service staff, who are charged with providing plastic straws and therefore judging whether a person should be entitled to this critical access tool. This is especially problematic when people may be regarded as 'not disabled enough' to deserve access, and where historically to identify in public as disabled has put people at risk. Furthermore, it risks exposing disabled people to further eco-ableism, as their plastic use is judged by others, as exemplified in scores of disabled experiences with the California straw ban. Like with many of the issues raised in this chapter, there is no easy alternative solution — but only because disabled people continue to be left out of decisions that affect our lives more than others. Leadership in these issues needs be from the disabled community; we need funding and backing from government and business to address the plastic pollution problem in a way that upholds disability rights. Being the master innovators that we are, we know we can find a solution if given the chance.

Taking disability-inclusive climate action

AMPLIFYING DISABLED CLIMATE NARRATIVES

At the time of writing, disability rights in climate action are in a state of flux. Much is happening behind the scenes, albeit at a slow pace, and there are essentially two mechanisms at work: the top levels of power, and grassroots activism. At the grassroots level there are plenty of climate leaders and activists with disabilities who are motivated by an awareness of just how dramatically climate change and climate action will affect them, and who are taking action at all levels — from the grassroots to diplomatic spaces. However, ableism prevents the media and the wider public from seeing these climate leaders as climate leaders or their disability identity is erased, which presents an added barrier to their voices being heard and seen. This has created a void of narrative in the space between grassroots campaigns and top levels of power, where perceptions must change in order for disabled people to be granted space and recognition as decision-makers and agents of change. One fantastic example of a climate leader who has bridged this space and is admired by protestors and decision-makers alike is Greta Thunberg. A young woman with Asperger syndrome, Thunberg is widely recognised as a leader on climate action and inspires youth around the world, disabled and non-disabled alike, including sparking the Fridays for the Future movement, which has rallied millions of youth to protest worldwide. Disabled people are already climate leaders; but the public is only just beginning to shift their perception of our community and create a space for us to participate meaningfully in demonstrations and consultation processes as our full, glorious disabled selves.

LEVERAGING INTERNATIONAL BODIES FOR ACCOUNTABILITY

In July 2020 during its 44th session, the United Nations Human Rights Council held the first ever dedicated meeting on people with disabilities in the context of climate change. The open panel event took place over video due to the Covid-19 pandemic, and included comments from New Zealand Disability Rights Commissioner Paula Tesoriero MNZM; the SustainedAbility Disability Climate Network was also able to make a formal statement via the Centre for International Environmental Law. The panel discussions aimed to promote and share knowledge on disability-inclusive climate action, increase understanding of the impacts of climate change on the rights of persons with disabilities, and highlight the benefits of disability-inclusive climate action (UNHRC, 2020). The panel and report released afterwards confirmed what disabled activists had been discussing for years: that our community would be disproportionately affected by climate change. This validation was an important acknowledgement, and is a promising indication that the UN's systems will create space and time for disabled leaders. While the UN may feel distant from the grassroots movements where many of our disabled community begin, multilateral climate negotiations are essential steps in setting a high bar for human rights and climate action. As Secretary-General Dag Hammarskjöld said over half a century ago, 'The United Nations was not created in order to bring us to heaven, but in order to save us from hell.' For disabled disability rights, one might think of the UN as a dam, a holding position that can prevent governments from using the climate crisis as an excuse to erode our rights, creating more space for our communities to organise and take essential climate action on our own terms.

SUPPORTING DISABLED-LED ACTIVISM

As co-founding members of the SustainedAbility Disability Climate Network, a growing network of disabled people working at the intersection of climate change and disability rights, we know that this is a critical step for our community. We are a network for disability-led climate action at all levels of governance, from grassroots activism to national government and international negotiations with the UN. We argue that Disability Justice is climate justice and disability rights must be upheld in climate action. Recently we have been privileged to take this mission to the UN systems in order to ensure that our rights are upheld in multilateral climate negotiations.

Every year — barring 2020 due to the Covid-19 pandemic — the United Nations Framework Convention on Climate Change (UNFCCC) holds international negotiations, with countries signed on to the framework tasked with figuring out a global approach to tackling climate change. These negotiations are called Conferences of the Parties (COPs) — where 'parties' are nation states — and have been going on for 26 years now. Non-governmental groups, organisations and civilians can participate in these COPs as 'constituencies' within civil society. Since 2017 we have attended these COPs, doing groundwork towards shifting the eco-ableist narratives in these spaces, building ally-ship and encouraging political will on disability and climate justice. After three years of hard work, we have succeeded in getting disability climate concerns on the radar: in a first for disability-related protest, we held a sanctioned action outside of a negotiation space calling for a disability constituency. The action gained much attention, from the press to negotiators alike, and we made some massive gains: other constituencies are supporting our case for greater disability inclusion and accessibility, and key decision-makers are beginning to understand the importance of disabled people's involvement in climate negotiations.

A disabled constituency to represent our rights in international climate negotiations on climate change is something we have been working towards, on the ground at the climate negotiations and within our communities, as well as calling for its establishment throughout the UN system — including at the 2018 Conference of States Parties to the Convention on the Rights of Persons with Disabilities during an intervention in the general assembly.

> *We [disabled people] need to be included in a meaningful capacity within the UNFCCC and its Conference of the Parties process as key stakeholders; without this we do not have representation to what will be our community's greatest challenge . . . we need a Person with Disability Climate Action Plan and we need it now . . . unless we take the role of leadership, climate change will leave all of us behind.*
> *Jason Boberg (11th session of the Conference of States Parties to the Convention on the Rights of Persons with Disabilities, June 2018)*

Our hope is that this work in the UNCRPD and UNFCCC will provide a template of best practice to accelerate disability climate action, leadership and governance across the world, including in Aotearoa, to make sure that we are not left behind in our country's climate responses but are instead leaders in upholding our rights and addressing climate change.

Closer to home, some of SustainedAbility's domestic work includes outreach, education, capacity building and making submissions to parliament and councils on proposals that will directly affect disabled people and our climate. Recently, for example, we made a submission on the New Zealand Transport Agency's (NZTA) 'Accessible Streets Regulatory Package', which aimed to set new rules on the use of roads and footpaths. The proposed package and consultation was anything but accessible and would allow cyclists to use the footpath with no speed limit, putting both disabled pedestrians and cyclists at risk. SustainedAbility composed an open letter, bringing together disabled organisations and climate groups such as The New Zealand Disabled Persons Assembly, Greenpeace New Zealand, 350 Aotearoa, Oxfam New Zealand and 350 Pacific. We asked that:

> *All infrastructure projects, including green infrastructure and transport, are fully accessible and uphold the rights of disabled people — including their participation and engagement in decision making and design. We urge the New Zealand government to respect its commitments under the United Nations Convention on the Rights of Persons with Disabilities when considering infrastructure development, 'shovel ready projects', and the Accessible Streets Package.*

The consultation and its recommendations had the unfortunate outcome of pitting cyclists against disabled people in an 'us versus them' debate that drifted far from the priority of people's safety and also sidelined the many disabled people who are also cyclists and deserve accessible active transport options. Working with the Disabled Persons Assembly NZ and allies in climate organisations, we gained more traction and support from the wider disability and climate activism communities, and eventually were pleased to see NZTA extending their consultation timeframe and committing to a disability-specific risk assessment. This is still under way, but is an optimistic early step in creating an urban environment that centres the rights of disabled people *and* helps reduce climate emissions. It also, importantly, proves the potential of disability organisations like SustainedAbility to create positive change in climate- and health-related issues, by bringing our unique perspectives as disabled people to policy issues that would otherwise entrench injustices and inaccessibility for disabled people and the wider community.

How you can be an ally for disability climate justice

So what can you do, as an individual?

- *Do your homework:* take the time to familiarise yourself with our movements by accessing the many resources available online to learn more about the history of disability rights as a movement and the issues raised in this chapter. The Disability Justice Primer by Sins Invalid (2019) is an excellent start.
- *Do work on dismantling ableism* in your personal thinking, actions and advocacy. This is ongoing and uncomfortable work, but it is critical that we all recognise how we play a part in upholding oppression, and work to change it. This is a necessary foundation for climate justice and action.
- *Do make strong relationships* with disabled people and our organisations within the climate movement.
- *Do preliminary research* before asking disabled people and organisations for one-on-one advice on how to improve your practices or your organisation's policies. It's important to take the first steps yourself, and remember that excessive requests for educational labour can divert our time, energy and resources

away from actions that centre our own communities.

- *Support disabled activists and organisations on our own terms*, making space for our own actions and platforms, through 'signal-boosting' or amplifying the messages, requests and profiles of disabled climate campaigners; and providing resources and financial support where you can.
- *Prioritise accessibility* if you're involved in organising protests or other events, and ask about accessibility if you are attending events that have not advertised this information.

It's important you remember that dismantling ableism is a life-long journey and we don't always get it right. We will all make mistakes, but crucial to the process is learning how to take critical feedback on board appropriately and learn from it to improve your practice in the future. We can't tackle climate change successfully without dismantling ableism, and dismantling ableism requires us all to be humble, and to be willing to be challenged.

We can't tackle climate change successfully without dismantling ableism, and dismantling ableism requires us all to be humble, and to be willing to be challenged.

Government action

Lastly, what can we demand of political action regarding climate change and disability in Aotearoa going forwards?

DISABILITY GOVERNANCE AND CONSULTATION IN CLIMATE BODIES

First, the Climate Change Commission needs a working group on disability. It already has a mandate to advise government on what is safe and for whom, and so it must ensure appropriate engagement with disabled people. This is a major opportunity to embed the rights and perspectives of our community from the outset, rather than relying on our community's advocacy and activism to fix issues after the fact. Similarly, local government needs to include disability rights and disabled people in their climate action plans, disaster-relief planning and the committees and bodies that make these decisions. At the time of writing, the Auckland Climate Plan does not mention disabled people once. From emergency exits to transport and climate shelters, councils have a responsibility to their disabled communities to consult on all climate matters that affect them. We have an opportunity to make transport and cities more inclusive and accessible, but instead we see the same mistakes being repeated, only now with electric engines. This reinforces the need for disabled people not just to be called on for answers to specific questions, but for local councils and our House of Representatives to have visible disabled representation from people with a connection to our community and our rights-based whakapapa.

MINISTRY FOR DISABLED PEOPLE

Finally, as 24 per cent of New Zealanders have a disability and as we face the ageing of our population, we must discuss the need for a Ministry of Disabled People and for that portfolio to be held in cabinet. At present we do not have a system that navigates both service and advocacy or that has the required political power to make effective reforms. Just as every mechanism within government must consider climate change, so too must they consider those who are disproportionately harmed by the effects of climate change. All climate action must involve disabled people, and ensure that our inclusion comes with real power for change. It is too late to not include the innovative, creative and insightful contributions from a quarter of our population. As much as our community needs urgent climate action, New Zealand as a whole needs our disabled wisdom.

INTERNATIONAL LEADERSHIP

If New Zealand wants to live up to its international reputation as a fair and progressive society, our government has work to do. There are many pressing needs for our disabled community, and for other marginalised communities throughout Aotearoa. However, the New Zealand government could go a long way to supporting disabled-led climate action by supporting calls for a Disabled Persons Constituency within the UNFCCC climate negotiations so that disabled people can participate fairly and safely.

Establishing a formal constituency will not guarantee disability rights on it's own, but it would facilitate further participation of disabled people — whether civil society activists, representatives of disabled people's organisations or state party representatives themselves — in these spaces where so many decisions that affect us are made without us. A formal constituency within the UNFCCC would enable disabled participants to come together from Aotearoa and around the world and find solidarity, and crucially connect and organise to raise the profile of disability and climate intersections in the spaces where decision-makers are most likely to listen and take action. Backing such an initiative could be as simple as our New Zealand government putting forward a statement, or sending an email to reiterate and support existing groundwork being done in this space — a small task with enormous potential impact for our community.

Conclusion

The New Zealand climate movement has historically excluded disabled people, both in direct lack of representation and through discrimination, ableist narratives and inaccessible events. In 2020, after many years of advocacy from disabled climate activists, New Zealand's leading climate organisations started to become aware of the intersection of disability rights and climate change, and some are taking action to better engage with disabled climate activists and the wider disability community. This is a promising start. However, more comprehensive and intersectional work is required to ensure that this knowledge and more inclusive practices make their way into the wider climate movement, ensuring that the movement is safe for disabled people and makes space for our leadership, communities and knowledges.

Having discussed previous models and movements within the disability rights and justice spheres, we recognise that the concept of Disability Justice, as set out by Sins Invalid, is an important and useful framework for effective, ethical and inclusive movement-building. Through Disability Justice we understand the need to address the root causes of climate change, which are also root causes of other social injustices, if we are to make any meaningful gains in preventing climate change, mitigating its worst effects, and equitably adapting to the changes already locked in. It is too late to depend on surface-level solutions — the urgency of climate change and the impacts it has and will continue to have on our communities necessitate bold action that tackles systemic drivers of climate change at the root. The authors hope that readers will be encouraged by this chapter to read *Skin, Tooth, and Bone: The Basis of Movement is Our People — A Disability Justice Primer* (Sins Invalid, 2019) to draw further on these concepts and incorporate them into their own climate activism and advocacy.

Finally, we hope that this chapter has provided a broad understanding not only of how climate change disproportionately affects our disabled community, but also of how integral diverse and disabled perspectives are effectively tackling this issue of our time. Our community has always connected, organised and innovated in the face of overwhelming challenges, and we know that we will continue to do so in the face of climate change. With the support of allies and accomplices committed to Disability Justice and climate justice, we can take real action on climate change and build a more sustainable, inclusive and joyful future where diverse communities thrive.

To finish with the words of perhaps the most well-known climate activist of our time, and a fellow member of the international disability community:

In some circumstances it can definitely be an advantage . . . to be neurodiverse, because that makes you different, that makes you think differently . . . especially in such a big crisis like this, when we need to think outside the box. We need to think outside our current system, we need people who think outside the box and who aren't like everyone else. (Thunberg, 2019)

References

J. G. Boberg. Climate change and disability rights, General Assembly Intervention at the 5th meeting of the 11th session of the Conference of States Parties to the Convention on the Rights of Persons with Disabilities, 12–14 June 2018.

R. Bhatia. 'Barrier after barrier' as disabled community locked out of housing. *Stuff*, 16 May 2019. https://www.stuff.co.nz/auckland/112717220/barrier-after-barrier-as-disabled-community-locked-out-of-housing, accessed 31 January 2021.

B. Flaws. Government plan to phase out single-use and problem plastics announced. *Stuff*, 12 August 2020. https://www.stuff.co.nz/business/green-business/122422225/government-plan-to-phase-out-singleuse-and-problem-plastics-announced, accessed 31 January 2021.

M. Green. How PG&E's power shutoffs sparked an East Bay disability rights campaign. KQED, 6 November 2019. https://www.kqed.org/news/11784435/how-pges-power-shutoffs-sparked-an-east-bay-disability-rights-campaign, accessed 31 January 2021.

S. Harrington. How extreme weather threatens people with disabilities. *Scientific American*, 18 September 2019. https://www.scientificamerican.com/article/how-extreme-weather-threatens-people-with-disabilities/, accessed 31 January 2021.

V. Ho. California power shutoff: how PG&E's action's hit the medically vulnerable the hardest. *Guardian*, 11 October 2019. https://www.theguardian.com/us-news/2019/oct/11/california-pge-utility-power-shutoff-disabled, accessed 31 January 2021.

L. Jackson. We Are the Original Lifehackers, May 30th 2018, New York Times. https://www.nytimes.com/2018/05/30/opinion/disability-design-lifehacks.html

J. Jambeck, R. Geyer, C. Wilcox, T. R. Siegler, M. Perryman, A. Andrady et al. Plastic waste inputs from land into the ocean. *Science*, 2015, vol. 347, no. 6223, pp. 768–771.

S. Lintern. Coronavirus: unlawful do not resuscitate orders imposed on people with learning disabilities. *Independent*, 13 June 2020. https://www.independent.co.uk/news/health/coronavirus-do-not-resuscitate-dnr-learning-disabilities-turning-point-a9561201.html, accessed 31 January 2021.

National Council on Disability. National disability policy: a progress report. Washington, DC, USA: National Council on Disability, 2017. https://ncd.gov/sites/default/files/NCD_A%20Progress%20Report_508.pdf, accessed 31 January 2021.

One News. Disabled man enduring 'very stressful process' as wait for accessible house to rent continues. Auckland: TVNZ, 8 December 2020. https://www.tvnz.co.nz/one-news/new-zealand/disabled-man-enduring-very-stressful-process-wait-accessible-house-rent-continues, accessed 31 January 2021.

M. Roth, J. Kailes & M. Marshall. Getting it wrong: an indictment with a blueprint for getting it right. Disability rights, obligations and responsibilities before, during and after disasters. USA: The Partnership for Inclusive Disaster Strategies, edition 1, May 2018. https://disasterstrategies.org/wp-content/uploads/2018/08/5-23-18_After_Action_Report_-_May__2018.pdf, accessed 31 January 2021.

C. Schulte. People with disabilities needed in fight against climate change: UN releases first report on disability rights in the context of climate action. Human Rights Watch, 298 May 2020. https://www.hrw.org/news/2020/05/28/people-disabilities-needed-fight-against-climate-change, accessed 31 January 2021.

Sins Invalid. *Skin, Tooth, and Bone: The basis of movement is our people*. USA: Sins Invalid, 2nd edition, 2019. https://www.sinsinvalid.org/disability-justice-primer, accessed 4 February 2021.

Statistics New Zealand. Disability survey: 2013. Wellington: Statistics New Zealand, 17 June 2014. https://www.stats.govt.nz/information-releases/disability-survey-2013, accessed 31 January 2021.

L. Ternouth. Waiheke buses go electric in Auckland first. In *Checkpoint*, Radio New Zealand, 10 November 2020. https://www.rnz.co.nz/national/programmes/checkpoint/audio/2018772186/waiheke-buses-go-electric-in-auckland-first, accessed 31 January 2021.

G. Thunberg. Iin 'Eye on Earth', *CBS This Morning*, CBS Entertainment Group, 11 September 2019. https://www.youtube.com/watch?v=BQ4rBLCpEeM&ab_channel=CBSThisMorning

J. Tucker, M. Cabanatuan & A. McBride. Tragic but familiar narrative in Camp Fire: Most victims were older, disabled. SFGate, 10 December 2018. https://www.sfgate.com/california-wildfires/article/Camp-Fire-victims-13450654.php, accessed 31 January 2021.

UNHRC (United Nations Human Rights Council). 44th Session Panel Discussion on the Rights of Persons with Disabilities in the Context of Climate Change, 8 July 2020.

United Nations. Convention on the Rights of Persons with Disabilities and Optional Protocol. United Nations, 2006. https://www.un.org/disabilities/documents/convention/convoptprot-e.pdf, accessed 31 January 2021.

UNStats (United Nations Statistics Division). United Nations Disability Statistics Programme. https://unstats.un.org/unsd/demographic-social/sconcerns/disability/index.cshtml

UNStats, 2021, accessed 10th February 2021.

A. Weibgen. The right to be rescued: disability justice in an age of disaster. *Yale Law Journal*, 2015, vol. 124, no. 7, pp. 2202–2679.

A. Wong. In California wildfires, disabled people may be left behind. Curbed San Francisco, 13 November 2018. https://sf.curbed.com/2018/11/13/18087964/califorina-wildfires-disabled-people-eldery-evacuation-disabilities, accessed 31 January 2021.

World Health Organization. World Report on Disability 2011. https://www.who.int/teams/noncommunicable-diseases/disability-and-rehabilitation/world-report-on-disability, accessed 31 January 2021.

C. Young. Visually impaired left out of talks as Christchurch switches to hard-to-see teal buses. Radio New Zealand, 11 December 2020. https://www.rnz.co.nz/news/national/432616/visually-impaired-left-out-of-talks-as-christchurch-switches-to-hard-to-see-teal-buses, accessed 31 January 2021.

E. Young. What it's like to experience a bushfire evacuation while living with a disability. SBS News, updated 8 January 2020. https://www.sbs.com.au/news/what-it-s-like-to-experience-a-bushfire-evacuation-while-living-with-a-disability, accessed 31 January 2021.

Taking matters into their own hands: Youth activism and climate change in New Zealand

SOPHIE HANDFORD
School Strike 4 Climate and Kāpiti Coast councillor

As we wake up to the fact that we are running out of time to safeguard a liveable planet for future generations, due to climate change, the need for change is increasingly relevant. The political and social changes required to address this critical issue are sadly lagging behind the rate of climate change itself.

We know we need to make some serious shifts in the way we operate and the things we value as outputs, from wealth to people and the planet. The report by the Intergovernmental Panel on Climate Change (IPCC) demonstrates that this shift must happen within the next 10 years, by which time we must halve the world's carbon emissions to have a chance at keeping temperature rise below 1.5°C. It also confirms that temperature rise must be limited to no more than 1.5°C if we are to avoid the catastrophic impacts of irreversible climate change.

Things have to change, and we have to transition our entire system. As supported by the IPCC report, this has to happen quickly, as each day we let slip by is another day closer to that 2030 deadline — another day closer to being certain we will have a liveable planet to pass on to future generations. Our political leaders seem to be slow to act, but there are many New Zealand citizens who fully understand our potential reality and are coming together to put pressure on those in positions of power to wake up.

The need for meaningful action on the climate crisis is becoming exponentially more pressing, so there are increasing numbers of citizen movements forming, growing and connecting. Their core demands and purpose are centred around raising the profile and prioritisation of climate change among the general public, our political leaders and those in positions of power such as CEOs of major companies. Many of these movements are using activism — vigorous campaigning to bring about political or social change — in an effort to achieve their collective aim.

In November 2018, as an 18-year-old recent high-school graduate from Kāpiti, I saw on Instagram videos and photos of the climate strike being led by the youth of Australia. I could see crowds of young people filling the streets, making their call for climate justice heard. The chants of thousands of voices, coupled with the emotions on the faces of those marching and the signs they held high in the air, resonated deeply with me. It connected with the fear I had been feeling, knowing the future my generation would inherit if we failed to treat climate change like the crisis that it is and act, centring on climate justice — the concept that not all people are equally affected by climate change and striving to address those inequities.

Just a few days later, I founded a New Zealand branch of School Strike 4 Climate (SS4C NZ). Initially I drew on my own high-school network, and many of my friends were extremely excited to be involved in the beginnings of this movement here. One of my teachers at Kāpiti College also added me to a Facebook group of young environmental activists from across the country, and I used this as an avenue to reach more people and get the movement off the ground.

Activism tends to be relatively fast-moving, and the creation of SS4C NZ was no exception. Within only three days, we had a social media presence and a published website. Isla Day, Bree Renwick, Francesca Griggs and I had a few video calls to plan for the weeks and months ahead. We worked with many different people and groups, including scientists, older activists and established climate groups, to put together a clear set of demands for elected leaders. We were also fielding all the sign-ups coming through our website. I decided to phone the first person who signed up from each town, to ask them if they'd be interested in becoming a regional coordinator for their community. The SS4C NZ movement quickly began to grow. The first big event SS4C NZ was involved with was the international strike for climate on 15 March 2019, followed by the 24 May and 27 September climate-strike events.

SS4C NZ is a student-led, student-powered movement made up of young New Zealanders from every corner of the country, mobilising to put pressure on the government to take immediate, transformative action on the climate crisis. Students involved range in age from 8 to 22, united by our shared concern for our futures, and our belief that achieving climate justice is and must be possible. In line with the international SS4C and Fridays For Future movements, our overarching demand is simple: our leaders must take urgent action on the climate crisis and protect our generation's, and all coming generations', right to a safe future on this planet.

SS4C NZ is a very young movement within a broad-spanning, diverse and well-established climate movement in New Zealand. Many different groups and communities have been part of weaving together the fabric of this movement, over many years. We recognise that our activism efforts are able to exist on the back of decades of resistance and mobilisation by those who have come before us and those who we continue to walk alongside now. We all take great inspiration

> **SS4C NZ is a very young movement within a broad-spanning, diverse and well-established climate movement in New Zealand.**

and guidance from these activists and communities and the changes New Zealand has already seen as a result of their relentless activism.

We acknowledge, for example, an important stream of the wider climate movement in New Zealand: the campaign to end all exploration and extraction of oil and gas in both our lands and our waters. In 2017, our newly elected coalition government announced the end of new offshore oil and gas exploration in Aotearoa New Zealand, off the back of decades of campaigning and activism led by various communities, non-governmental organisations such as Greenpeace New Zealand, and tangata whenua. This was a very significant moment for the climate movement in New Zealand, and will be remembered as a stellar example of the power of grassroots, people-powered movements and activism.

Despite this world-leading move to end new exploration and extraction, there are still so many processes, structures and core values that need to change within our current system if we want to truly tackle the climate crisis and build a more equitable world for coming generations. SS4C NZ's focus is on ensuring the government knows the impact its decisions have on the younger generation, because they hold in their hands so much of the power to determine what the future looks like for generations to come. We need bold, transformative and urgent action on a systemic level to ensure that this future is livable and that our tamariki are able to continue to connect with our beautiful natural environment.

Holding this focus at the centre of our work and activism means that we remain committed to the cause through the challenges that being involved in the world of activism can pose. We created a team of people who all had this same 'Why', allowing us to work together well and create a movement which fosters a strong sense of kindness, collaboration and connection.

Working from this space allowed us to consider different ways that our activism could be most effective. Our main tactic so far has been mass mobilisation, through striking from school for climate justice. We have also created other ways for people to be involved in the movement, recognising that this is crucial if we want the movement to be inclusive.

People in our team have contributed to this mass mobilisation effort in very different ways. Those who are passionate about art have used their talents to mobilise members of their community by connecting with people on a visual level. We have had young people on the team who are incredible writers, social media experts, speakers, musicians and videographers. We consciously try to harness the passion, skills and energy of each of our members.

Through creating an environment where people feel empowered to use their skills in a team, we can see a collective action also becoming a way for young people to take individual action. The act of being involved in an activism movement or being part of a collective is also a great way for people to begin their journey into taking action in their own lives, as ideas can be shared between members of these groups and movements and also through the wider climate movement.

The youngest of our coordinators are 8, 10 and 12 years old, which demonstrates how even tamariki this young are feeling like they have to dedicate their childhood years to standing up for their right to a future. They are shouldering a burden that they ultimately shouldn't have to bear. This has the potential to increase anxiety in our young people and can bring a real feeling of powerlessness when they come up against obstacles or negative opinions on their activism journey. The movement does, however, give us as youth an avenue to have our voice heard and to make a difference, which hopefully in turn can decrease feelings of fear and anxiety.

Young people are becoming more and more prominent in this space right across Aotearoa New Zealand

Young people are becoming more and more prominent in this space right across Aotearoa New Zealand, both leading and getting involved with activist movements. We take great guidance and strength from other youth activist groups, particularly Indigenous youth groups, who have been working alongside their communities on the issue of climate change for many years.

Two important groups are the Pacific Climate Warriors, a group of Pasifika youth who have been active in Aotearoa New Zealand for the past few years, and in the wider Pacific region for over a decade; and Te Ara Whatu, a collective of Māori and Pacific young people from Aotearoa who have sent Indigenous youth delegations to the international UN Climate Conference since 2017. They were the first all-Indigenous youth delegation from Aotearoa to attend these negotiations. Both the Pacific Climate Warriors and Te Ara Whatu work tirelessly to bring Indigenous knowledge and understanding to spaces which are often otherwise dominated by western scientific frameworks and world views. They also fight on the front lines for their communities here in Aotearoa New Zealand and the Pacific, running campaigns and events and creating space for their local communities to come together and add their voice to the call. These communities are not only disproportionately affected by the impacts of the climate crisis, but through their Indigenous knowledge and understanding they also hold the solutions that our world so desperately needs. For these reasons, it is fundamental that these voices are at the centre of conversations about the climate crisis, and achieving climate justice. It is crucial that the activism of these groups and others representing the same kaupapa is amplified.

The climate activism movement in New Zealand has a real sense of intergenerational strength and kindness extending which crosses the perceived boundaries of generations.

We are seeing that climate change is an issue that increasing numbers of young people are feeling connected to and have a drive to do something about. Whether or not we take action on the climate crisis directly determines what future we will be inheriting. This can feel like a scary prospect, but it is something we have the power to influence. Movements such as SS4C, Pacific Climate Warriors, Te Ara Whatu and others are giving these young people a way to be involved in this change and find their place and their voice.

The climate activism movement in New Zealand has a real sense of intergenerational strength and kindness extending which crosses the perceived boundaries of generations. As young people entering into the climate-activism scene, there has been so much to learn, and we continue to constantly learn on the go. Having that shared knowledge and experience flow between generations has strengthened the movement, giving it greater depth and ability to challenge those in power to take the required action to mitigate the climate crisis.

S eptember 20 and 27 of 2020 were declared two international days of action for climate by the International Fridays for Future Committee. On these days, each country involved in the international movement was encouraged to organise something which demonstrated the importance and power of intergenerational solidarity. We decided to organise an intergenerational climate strike, so began building connections with unions, businesses, Rotary and other groups we hadn't yet reached out to, due to the earlier strikes being more specifically around how we could mobilise youth. The response was overwhelmingly positive, and the speed at which intergenerational support developed for the climate strike was incredible. We had 290 businesses pledge their support within four days, under the umbrella of the Not Business As Usual Campaign. Although SS4C is a youth-led movement, we see great strength and power in having people of all ages and backgrounds walking alongside each other in this collective mission.

Activism provides a valuable space for people to find their voice when it can otherwise be easy to feel powerless, paralysed by fear or by the scale of the climate crisis. The realisation that you are not in this alone and that there are thousands of others thinking the same thoughts and wanting the same action is a feeling like no other. It is empowering and life-giving, and can propel us forward into doing something tangible together to campaign and put pressure on our leaders.

I felt this way when first founding SS4C NZ. I felt that whatever I did would not make a difference in the big scheme of things, considering that globally just 100 companies emit around 70 per cent of the world's total greenhouse gas emissions. Talking with some of my friends about this feeling, I found they had very similar thoughts. But we were incredibly inspired by Greta Thunberg and her initiative of sitting outside the Swedish Parliament to protest the inaction of her leaders. That's when we decided to put our heads together and try to create a movement here that provided an avenue for youth across New Zealand to find their voice in the conversation around our collective future. And when people find their voice, they find their power to act.

Through the examples mentioned, and the movements which continue to emerge, the power of people is clear. When a group of people come together with a goal, vision or clear purpose and understand how to communicate these values, their activism has the potential to make really important change for the future of Aotearoa New Zealand.

With the window of time we have to make this change closing day by day, we are seeing activism in an array of different forms, offering a variety of different levels at which people can be involved, depending on how open they are to radical activism, non-violent direct activism and other forms of peaceful but vigorous protest.

With the window of time we have to make this change closing day by day, we are seeing activism in an array of different forms.

I have learnt that vigorous campaigning can look very different in a variety of contexts, depending on people's skill-sets and passions and the group they are trying to influence, but also that there are some things which remain constant. These include the need to know what you are campaigning for and have all messages united behind this one aim; the need to know how to bring others on board and grow the movement; how to integrate what's important to each of us into our activism and then communicate this to others; and how to honour Te Tiriti o Waitangi throughout the process, working alongside tangata whenua.

When all of these things come together, activism can foster whānau-like bonds and create a support network for people working in this space, enabling them to feel safe and valued when they express how they might be feeling about the current situation. The sense of community that is created through activism only strengthens our collective resilience as we move forward into the future together.

The climate crisis is too important for us to not act on. Everything is at stake, so we must do all that we can, using activism as a tool and other levers too, such as using our vote for the planet and having conversations with friends and family about the importance of these next 10 years. As SS4C team member Ella Magnusson said when addressing the Wellington City Council in July 2019, 'When an animal is hurt, it doesn't just die. It keeps fighting until either its wound is healed or until it dies. I am going to be fighting this until either we tackle the climate crisis or until I die. It is too important, and if we want to have a planet to pass on to generations to come, this is what we must do.'

PART 3

The Solutions

Bigger than me? What individuals can do to combat climate change

JAMIE MORTON
Science reporter, *The New Zealand Herald*

New Zealand's climate change outlook is crammed with big, scary numbers. By the middle of the century, when students who took part in 2019's climate marches will be in their 40s, with children of their own, average temperatures will be, at best, 0.7°C warmer than today.[1] That means more heatwaves, droughts and floods.

By the end of the century, assuming that greenhouse gas emissions continue to climb close to current levels, many parts of our country will record more than 80 days a year above 25°C.[2] Most places typically only have between 20 and 40 days a year above that level now, yet already about 14 elderly people in Auckland and Christchurch die each year when the mercury climbs above only 20°C. If global temperatures climb just 1, 2 or 3°C above current levels, that same death rate could rise to 28, 51 and 88 per year respectively.[3]

The best-case scenario is that global warming stops somewhere between 2050 and 2070. The worst case is that, by 2100, New Zealand temperatures end up more than 3°C above present — and sea levels rise 1 metre or even higher.[4] Heavy rainfall and flooding events would quadruple, and extreme high temperatures could make many mid-latitude areas virtually unliveable in summer.

But perhaps the bleakest statistic of all is this one: 60 per cent.[5] That's the proportion of Kiwis, as indicated by a 2018 Ipsos poll commissioned by insurer IAG, who aren't sure whether humanity will

1   'Climate Change Scenarios for New Zealand', NIWA, https://niwa.co.nz/our-science/climate/information-and-resources/clivar/scenarios#:~:text=Higher%20elevation%20warming%20particularly%20marked%20for%20maximum%20temperature.&text=Decrease%20in%20cold%20nights%20(minimum,90%25%20%5B8.5%5D%20decrease.
2   'Climate Change Implications for New Zealand', Royal Society Te Aparangi, 2016, https://www.royalsociety.org.nz/what-we-do/our-expert-advice/all-expert-advice-papers/climate-change-implications-for-new-zealand/
3   'Human Health Impacts of Climate Change in New Zealand', Te Aparangi Royal Society, October 2017, https://www.royalsociety.org.nz/assets/documents/Report-Human-Health-Impacts-of-Climate-Change-for-New-Zealand-Oct-2017.pdf
4   New Zealand Sea Rise, https://www.searise.nz/
5   'IAG-Ipsos poll: Kiwis pessimistic that we will meet climate change challenge', July 2018, https://www.iag.co.nz/latest-news/articles/IAG-Ipsos-poll-Kiwis-pessimistic-that-we-will-meet-the-challenge-of-climate-change.html

do what's needed to avoid the worst of climate change. Only one in ten are optimistic that the world will act in time.

Another Ipsos poll[6] suggests that just half of New Zealanders think that taking our own steps will alter the impact climate change ultimately has on us.

Are they right? As opponents to ambitious climate action like to remind us, New Zealand contributes just a tiny fraction — about 0.2 per cent — of the world's total greenhouse gas emissions. Therefore we might say our climate change fate won't be dependent on us giving up red meat or buying an electric car, but on big-polluting nations like China, the US and India succeeding or failing to arrest their emissions curves before we blow through the all-crucial 2°C mark.

But equally, we might argue that if none of us takes action, a wetter, warmer, wilder future of the grimmest projections really is inevitable. Perhaps Kiwis get this, too: most of us want our government to be bold, even if the rest of the world doesn't seem to bother.[7]

And there's much room for improvement. On a per capita basis, New Zealand has an embarrassingly large carbon footprint, emitting some 18 tonnes of greenhouse gases per person each year.[8] By that measure, we're near the top 20 countries in the world — and within the top five in the OECD.

As climate campaigner Greta Thunberg puts it: you're never too small to make a difference.

---

6   'IAG-Ipsos poll: Kiwi concern grows about climate change', July 2019, https://www.scoop. co.nz/stories/PO1907/S00330/iag-ipsos-poll-kiwi-concern-grows-about-climate-change. htm

7   'Poll: Few Kiwis think the world will overcome climate change', *The New Zealand Herald*, 15 July 2018, https://www.nzherald.co.nz/nz/poll-few-kiwis-think-world-will-overcome-climate-challenge/F33CDULT7K47LQDYZI4ZRDUCFU/

8   'New Zealand's out-sized climate change contribution', *Stuff*, December 2018, https://www. stuff.co.nz/environment/climate-news/109053475/new-zealands-outsized-climate-change-contribution

The home front

So, where to start? Look no further than where you live — and shop. Statistics New Zealand figures from late 2020[9] tell us that households are the largest contributor to New Zealand's carbon footprint — making up 71 per cent of it — because of the goods and services we consume. That includes trips in the family cars, food and beverages, and the way we use power and water.

Still, the carbon footprint of an average Kiwi home is a tough one to calculate. Who lives in a household and how much they spend, for instance, can explain nearly three-quarters of the variation that researchers have found in emissions.

One frequently referenced paper, by Wellington-based think-tank Motu Economic and Public Policy Research,[10] points out food-related emissions account for roughly 40 per cent of what households produce, no matter their income. Utilities like power make up nearly a third of emissions for poorer households, compared with just over 20 per cent for the wealthiest ones. Yet richer homes still have a larger carbon footprint, partly because they consume more items such as meat and dairy, and they burn through much more petrol and diesel.

Fortunately, all of this points us towards some reasonably fast and easy gains we can make, starting right now. If every Kiwi didn't drive one day a week, switched off their appliances at the wall, and converted to low-energy lightbulbs, New Zealand could save 386,500 tonnes of CO^2-equivalent greenhouse gases per year[11] — or around 5 per cent of our total emissions.

---

9   'Transport drives households' carbon footprint up', Stats NZ, August 2020, https://www. stats.govt.nz/news/transport-drives-households-carbon-footprint-up#:~:text=Transport%20 accounted%20for%2037%20percent,of%20electricity%2C%20contributed%2012%20percent.

10 'Who's Going Green? Decomposing the Change in Household Consumption Emissions 2006–2012', Motu Economic and Public Policy Research, https://motu.nz/our-work/environment-and-resources/emission-mitigation/shaping-new-zealands-low-emissions-future/whos-going-green-decomposing-the-change-in-household-consumption-emissions-2006-2012/

11 WWF New Zealand, https://www.wwf.org.nz/what_we_do/climateaction/reduce_your_impact

We can insulate our homes and install solar panels. We can try having shorter showers, or outside the peak hours between 4 p.m. and 8 p.m., when there is the greatest demand on the grid for fossil-fuelled electricity generation. Even a more efficient showerhead can make a difference: with one of these, a person who showers once a day can save the equivalent of burning a 20 kg bag of coal in a year.

We can wash our clothes in cold water, make items such as clothes and home appliances last, buy second-hand, and repair rather than replace. We can buy food that's sourced as locally as possible, buying New Zealand-grown rather than imported fruit, for example.

We can avoid products with too much packaging, and compost our kitchen scraps and garden waste. We might go a step further and grow what we eat.

We use gadgets like Fitbits to track our exercise. How can we do the same with emissions? Researchers at Motu have put their knowledge into a handy calculator, called the Household Climate Action Tool (see https://insights.nzherald.co.nz/article/climate-action-tool). There's also the nifty CoGo app, developed by Kiwi entrepreneur Ben Gleisner, which goes further by analysing household spending to calculate your real-time carbon footprint and encourage you to lower it.

In the UK, where CoGo been rolled out ahead of its planned arrival in New Zealand, users initially learnt that they were typically churning through 1000 kg of CO^2 equivalent each month.[12] That is roughly seven times higher than where Britain needs to be by 2030 to keep its emissions within goals to limit warming to 1.5°C.

In other cases, we can make a difference by being more aware not only of *what* we're spending, but also whether we're spending on goods that are sustainably made. That's not always straightforward.

---

12   'Blown your budget? New app shows climate impact of your spending', *Stuff*, 23 June 2020, https://www.stuff.co.nz/environment/climate-news/300039721/blown-your-budget-new-app-shows-climate-impact-of-your-spending

According to Consumer New Zealand,[13] only half of shoppers have trust in environmental product claims, while nearly three-quarters struggle to work out which products are actually greener. There's a clear mandate for more clarity — and the New Zealand Sustainable Business Council reports that this matters to nearly nine in ten consumers.[14]

While Kiwis think electricity retailers and supermarkets are doing the most to reduce their footprint, they want brands in all industries to be more open and upfront about the steps they're taking — and actively communicate them.

Beyond shopping at outlets and websites that are based on eco-friendly principles — and there are now many to choose from — consumers can look out for product labels such the government-backed Environmental Choice tick, or AsureQuality or BioGro's organic certifications.

Climate action on a plate

Should we go further and take meat out of our diets? Let's look at what we know.

Today, agriculture covers nearly 40 per cent of global land, making agroecosystems the largest terrestrial ecosystems on the planet. Food production is responsible for up to 30 per cent of global greenhouse gas emissions, and nearly three-quarters of freshwater use.

As we in New Zealand know all too well, land conversion for food production is the single most important driver of biodiversity loss. Ditching animal protein is seen by an increasing number of people as the only way to deal with the fact that, by 2050, the world's population will

---

13   'Environmental concerns spur change in shopping habits', *Consumer NZ*, 5 February 2020, https://www.consumer.org.nz/articles/eco-concerns-spur-change-in-consumers-shopping-habits

14   'The influence of sustainability on New Zealand consumers', Sustainable Business Council, November 2019, https://www.sbc.org.nz/news/2019/the-influence-of-sustainability-on-new-zealand-consumers

hit 10 billion, rendering the demand for meat higher than the industry's ability to supply it.

As prominent New Zealand science communicator Associate Professor Siouxsie Wiles points out, studies suggest that climate change is going to lower the yields and nutritional value of staple crops like corn and wheat, expand the areas where crop pests can survive, and make it more difficult for farmhands to work at certain times of day due to the heat.

'In other words,' she says, 'we simply can't rely on our current land-hungry, water-thirsty, pollution-heavy and extinction-inducing ways of producing food if we are to feed the ever-growing human population as our environment changes around us.'

Science is also increasingly telling us about the health benefits of a more climate-friendly diet, consisting of less red meat. Otago University researchers have found that eating less red meat could be key to New Zealand not only significantly slashing emissions but also saving billions of healthcare dollars over coming decades.[15] Specifically, they showed a population-level shift to diets rich in plant foods such as vegetables, fruits, whole grains and legumes could — depending on the extent of changes made — cut diet-related emissions by between 4 and 42 per cent annually.

More strikingly, if all Kiwis adopted an exclusively plant-based diet tomorrow, and avoided wasting food unnecessarily, we'd achieve what would be equivalent to a 60 per cent drop in emissions from cars. As a bonus, Kiwis could collectively enjoy up to 1.5 million more 'life

15   'Plant-based diet: The case to help slash emissions and save billions', The New Zealand Herald, 23 January 2020, https://www.nzherald.co.nz/nz/news/article.cfm?c_id=1&objectid=12302611

years' — that's those equivalent to a year of optimal health — and save our health system between $14 billion and $20 billion over the lifetime of our current population.

There are already plenty of signs that a green shift is happening. By 2016, the proportion of Kiwis who stated that all — or almost all — of the food they ate was vegetarian had grown by nearly a third from four years earlier.[16] The sharpest rises came from among 14- to 34-year-olds, North Islanders and, perhaps surprisingly, men.

More-recent polling by Colmar Brunton indicates that about one in ten of us is now largely shunning meat, amid a growing shift to sustainable lifestyles.[17] Industry data similarly indicate a downward trend of red-meat consumption in New Zealand over the past 10 years, with beef, lamb and mutton down 38 per cent, 45 per cent and 72 per cent respectively.[18]

Rates of vegetarianism do tend to drop away among Kiwis in their 30s and 40s — likely because parents with kids to feed find it tougher to stay meat-free. However, veganism, once considered a radical-fringe movement, is also on the rise. An intriguing study by online cookbook Chef's Pencil on the countries and cities where veganism was most popular put New Zealand — an agricultural nation which we might think of as a carnivore's paradise — in third place.[19] The online cookbook used Google Trends to analyse the search-interest level — trawling for terms like 'veganism' and 'vegan restaurants' — across the world.

Supermarket chains have certainly noticed this trend, and have responded by stocking their shelves with more vegan, vegetarian and flexitarian (people who are primarily vegetarian but occasionally eat meat or fish). You'll now find vegan chewing gum and vegan 'fried

16   'Vegetarianism on the rise in New Zealand', Roy Morgan, 2016, http://www.roymorgan.com/findings/6663-vegetarians-on-the-rise-in-new-zealand-june-2015-201602080028

17   Colmar Brunton 'Better Futures' report, February, 2019, https://www.colmarbrunton.co.nz/better-futures-climate-change-concern-rising-but-plastics-top-of-mind-for-kiwis/

18   Beef + Lamb New Zealand: EAT-Lancet Report FAQs, https://beeflambnz.com/sites/default/files/news-docs/Beef%20Lamb%20New%20Zealand%20EAT-Lancet%20Report%20FAQs.pdf

19   'The Most Popular Countries and Cities for Vegans in 2018', Chef's Pencil, https://www.chefspencil.com/where-veganism-is-most-popular-around-the-world-in-2018/

eggs', along with plant-based sausages, hot dogs, patties and mince, and curious items such as 'bacon-style rashers' and 'chicken-free chicken'.

In terms of sustainability, we do need to consider whether it is fair to compare New Zealand's meat sector with its overseas counterparts. Beef + Lamb New Zealand argues that our farming systems, which make efficient use of land unsuitable for horticulture or arable production, are often unfairly compared with grain-fed, factory-farming models seen in the US.

Nearly half of our emissions come from agriculture . . .

Nearly half of our emissions come from agriculture — the bulk of that being methane from ruminant animals such as cows and sheep — and some farming models like intensive dairying do generally emit more greenhouse gases.

But the footprint of the sheep and beef industry has a different profile. Sheep and beef emissions have fallen by a third since 1990, in step with falling stock numbers, and, with some 2.8 million ha of forest on its land, the industry holds the largest collection of native bush outside the conservation estate, bringing carbon-offsetting benefits.

Can we move greener?

In 2018, prominent New Zealand physicist Professor Shaun Hendy began a one-man campaign that aimed to make us think about how we get from one place to the next. Hendy was inspired by a talk by Professor Quentin Atkinson, a fellow Auckland University researcher and an expert in how cultures change and evolve, who explained why we believe things even when there is no evidence, or the evidence is against us.

'He made the point [that] we often put faith in people who make sacrifices that demonstrate the strength of their conviction,' Hendy says. 'People who walk the talk can be more convincing than those who just talk.'

So he started walking the walk — by giving up flying for a year.

That meant travelling to Wellington by train and heading back to Auckland by overnight bus, and swapping flights to international conferences and meetings for Zoom calls. By the end of the year, he calculated he'd saved an equivalent 18 tonnes of carbon emissions.

Before the Covid-19 pandemic, it was estimated that aviation contributed to about 3.5 per cent of human-driven emissions.[20] It's also been projected that, as at 2020, the impact would have grown 70 per cent from 2005 levels.

That doesn't change the fact that Kiwis love to — and quite often need to — hop on a plane. In 2017, for instance, New Zealand residents departed on 2.83 million trips overseas, up 271,800 on the year before.[21] When our largest export industry — tourism — recovers from the Covid-19 crisis, it will return to its reliance on international travel. The $39 billion it reaped in 2018, after all, wouldn't have been possible had not 3.82 million visitors arrived on our shores.

That's not to say that aviation is the only part of the transport sector struggling to rein in its CO^2 pollution. New Zealand's gross emissions jumped 2.2 per cent between 2016 and 2017 — and one of the biggest increases was the 6 per cent bump (an equivalent of 863 tonnes of CO^2) chalked up by road transport over that period.[22] In all, transport makes up 20 per cent of the country's total emissions, and two-thirds of that comes from the cars, SUVs and utes we drive.

After repeated recommendations to adopt a 'feebate' scheme to

---

20   'Aviation contributes 3.5% to the drivers of climate change that stem from humans', University of Reading, *Phys Org*, https://phys.org/news/2020-09-aviation-contributes-drivers-climate-stem.html#:~:text=Aviation%20has%20been%20calculated%20to,climate%20change%2C%20new%20research%20shows.&text=The%20findings%20show%20that%20two,the%20rest%20from%20CO2.

21   International travel and migration: October 2017, Stats NZ, https://www.stats.govt.nz/tereo/information-releases/international-travel-and-migration-october-2017

22   New Zealand's Greenhouse Gas Inventory 1990–2017, Ministry for the Environment, https://www.mfe.govt.nz/sites/default/files/media/Climate%20Change/snapshot-nzs-greenhouse-gas-inventory-1990-2017.pdf

clean up our ageing vehicle fleet, the government has moved to bring in a Clean Car Standard, requiring importers to reduce the average emissions of the vehicles they are bringing into the country. But it won't apply to the rest of our existing fleet, which is one of the oldest in the developed world.

Why is age a factor? It's estimated that the average car drives some 12,000 km a year. Before 2008, that average car emitted around 215 g of CO^2 per kilometre (gCO^2/km). With more modern and efficient vehicles on our roads, that's dropped to 180 gCO^2/km — but there's still a long way to go.[23]

In our largest city, nearly four in ten Aucklanders now cycle regularly, while more than two-thirds frequently walk.

Atkinson himself argues that tackling New Zealand's transport emissions, and those from everywhere else, shouldn't be thought of as a sacrifice.

'This framing blinds us to the fact that, almost invariably, for everything we are asked to give up, there exist equally good, perhaps better alternatives that we likely haven't even considered,' he says. 'We confuse new opportunities for sacrifice.'

What are those new opportunities? Just as with cutting down on red meat consumption, the obvious one is for our health. There's plenty of research to show that trading a car trip for walking or cycling leads to longer, healthier lives, and happier, more social ones, too.

In our largest city, nearly four in ten Aucklanders now cycle regularly, while more than two-thirds frequently walk.[24] With more people taking up active transport, and better infrastructure helping people feel more confident on a bike, numbers are growing year on year.

23   'Clean Car "feebate" explained', *Consumer NZ*, August 2019, https://www.consumer.org.nz/articles/clean-car-feebate-explained

24   'Measuring and growing active modes of transport in Auckland', Auckland Transport, 2018, https://at.govt.nz/media/1977266/tra_at_activemodes_publicrelease-1.pdf

So, too, are the numbers of Kiwis buying electric vehicles. There are now more than 22,000 EVs on our roads, most of them in our major cities — though this is well short of the target of 64,000 EVs the previous National-led government set for 2021.[25]

That's partly because the upfront cost of a brand-new EV can be high: the cheapest option, an MG ZS EV, sits at around $50,000, but older models of Nissan's popular Leaf can still be found second-hand for around $10,000. And the benefits are many. EVs are far cheaper to run than petrol or diesel vehicles, and New Zealand's high renewable-energy levels mean their carbon-cutting benefits are much greater than in other countries, producing about 80 per cent fewer emissions than a traditional car with an internal combustion engine. Still, EVs have a perception problem to overcome, with Kiwi men apparently worried that owning one might dent their 'macho' image. A 2018 poll of Kiwi EV owners found the powerful thrum of an internal combustion engine was apparently more manly than the calm quiet of an electric engine — something more likely to appeal to women, apparently.[26]

As far as Joe Camuso of citizen-science project Flip the Fleet is concerned, the pros of EVs trump any negative stereotypes about them.

'Put a man or a woman behind the wheel for a test drive, and they are sold in a minute. It's a fast, quiet and comfortable ride,' he says. 'Later, you can fill in that they are also good for your purse and the planet.'

---

25   Monthly electric and hybrid light vehicle registrations, Ministry of Transport, September 2020, https://www.transport.govt.nz/mot-resources/vehicle-fleet-statistics/httpswww-transport-govt-nzmot-resourcesvehicle-fleet-statisticshttpswww-transport-govt-nzmot-resourcesvehicle-fleet-statisticsmonthly-electric-and-hybrid-light-vehicle-registrations/
26   'Electric cars not "macho" enough for some men — survey', *Otago Daily Times*, April 2018, https://www.odt.co.nz/news/national/electric-cars-not-macho-enough-some-men-survey

The people problem

Being more mindful of what we eat and buy, how much power we use and how often we fly or drive is one thing. But one climate-driven movement is taking action to a new extreme: opting against having children.

One small but headline-grabbing group based in the UK, dubbed BirthStrike, describes itself as a 'radical acknowledgement that our planet has entered a 6th Mass Extinction event due to man-made impacts on the environment'. The group points out that it respects other people's wishes to have children, and isn't calling for population control at the expense of climate action, but is merely taking the step itself.

So does family size matter? Scientists indeed say human population growth is intangibly linked to the mass-extinction event the group mentions. In fact, it's one of the largest ever recorded in the Earth's history, threatening a million species of plants and animals, and often discussed by heavyweight naturalists such as Dame Jane Goodall and Sir David Attenborough. The idea gained further weight when influential US congresswoman Alexandria Ocasio-Cortez told her Instagram followers in February 2019: 'There's a scientific consensus that the lives of children are going to be very difficult — is it still ok to have children?'

In the last major assessment by the United Nations' Intergovernmental Panel on Climate Change (IPCC), it was estimated that global carbon dioxide emissions might be lowered by nearly a third if contraception was available to all women who expressed a need for it.[27] Another study, published by Sweden's University of Lund and Canada's University of British Columbia in 2017, found that the single most effective measure an individual in the developed world could take

---

27   AR5 Synthesis Report: Climate Change 2014, UN Intergovernmental Panel on Climate Change, https://www.ipcc.ch/report/ar5/syr/

to cut their carbon emissions over the long term was to have one less child.[28] In fact, their study established that this step was 25 times more effective than the next most effective measure — living without a car.

Dramatically, a group of environmental scientists have argued that societies should embrace population ageing, as is being experienced here, and even decline. They cited multiple reports of the socio-economic and environmental benefits of population ageing, while pointing out that smaller populations made for more sustainable societies.

That sentiment also rings true in the IPCC's most recent report, which warns that high population growth will be a 'key impediment' to hitting the critical target of limiting global warming to 1.5°C.[29]

> **Dramatically, a group of environmental scientists have argued that societies should embrace population ageing, as is being experienced here, and even decline.**

Those nations with massive populations, such as India and China, are among the most significant contributors to climate change overall, despite relatively low impacts from each individual. While India and China have reasonably low population growth, it's expected that people born today in countries whose populations are still expanding rapidly will have a climate impact for generations to come.

It might be argued that slowing population growth is already happening here in New Zealand — but not for environmental reasons. Population growth since 2013 has been dominated by net migration, rather than the number of births, which, before the Covid-19 pandemic, had been running relatively steady at about 60,000 a year despite a

---

28   'Serious about stopping climate change? Have one less child, UBC study says', *National Post*, July 2017, https://nationalpost.com/news/canada/want-to-stop-climate-change-have-one-less-child-ubc-study-says

29   IPCC special report on the impacts of global warming of 1.5C, 2018, https://www.ipcc.ch/sr15/

decline in birth rates.[30] In other words, the number of births for every 1000 people is falling — but the growing population means *total* births remain at relatively high levels, reaching a recent peak of almost 65,000 per year in the period from 2007 to 2010. However, New Zealand's total fertility rate in 2017 was down to 1.8 births per woman, its lowest recorded level.

In any case, many environmentalists argue that focusing too much on population distracts us from tackling the root causes of the ecological crisis we've created. The global network behind the most commonly used indicator of our impact on the world, the ecological footprint, calculates that humans are chewing up natural resources about 1.7 times faster than they can be regenerated.[31] Even if everyone lived like people in supposedly 'clean and green' New Zealand, which in 2012 had the thirty-first-highest ecological footprint out of 188 countries, we'd need about 2.8 Earths to sustain our consumption.

As climate change minister James Shaw points out, 18 countries managed to slash their emissions between 2005 and 2015, even while their populations and economies grew over the same period of time.

'So while there's obviously a correlation between population growth and use of resources, particularly in wealthier countries which, like New Zealand, have the highest per capita emissions, the evidence shows that it is possible to decouple emissions growth from population and economic growth,' he says. 'The main challenge lies in adopting new technologies and business models and in being far more efficient with resources, which New Zealand has, so far, been slow to get started on.'

Put another way: people might be the problem, but we're also the solution.

------

30   Population projections overview, Stats NZ, http://archive.stats.govt.nz/browse_for_stats/
      population/estimates_and_projections/projections-overview/nat-pop-proj.aspx#gsc.tab=0
31   'Earth Overshoot Day 2018 is August 1', Footprint Network https://www.footprintnetwork.
      org/2018/07/23/earth-overshoot-day-2018-is-august-1-the-earliest-date-since-ecological-
      overshoot-started-in-the-early-1970s-2/

Advocacy and apathy

Shaw's argument speaks to a larger and arguably more important point: by focusing too much on individual action, do we risk shifting the blame from those bigger polluters?

Climate change campaigner David Tong says the choices we make as people and consumers are shaped by the fundamental economic and political structures we live in, and those structures still don't fully factor-in the price of carbon. That means that individual lifestyle or purchasing impacts are limited.

There are many ways people can lower their impacts, he argues, but we need to make sure we don't guilt-trip them or shift blame onto individuals. After all, more than a third of global emissions since 1965 can be traced to just the 20 biggest fossil-fuel companies,[32] and almost 70 per cent of global emissions can be tied to just 100 companies.[33]

That's not to say that legitimate carbon-cutting efforts aren't being made by big corporates here — notably with the newly launched Climate Leaders Coalition with its heavyweight members like Fonterra, Z Energy and The Warehouse Group — or overseas. Coca-Cola is trying to shrink its carbon footprint by a quarter within the next five years, the same timeframe within which McDonald's also aims to source all of its packaging from recycled materials, and beauty giant L'Oréal wants to become carbon-neutral.[34]

But should the same pressure be heaped on individual consumers? Tong feels that individual people have the most power not in lifestyle or purchasing decisions, but in compelling governments and companies to

32   'Revealed: the 20 firms behind a third of all carbon emissions', *The Guardian*, October 2019, https://www.theguardian.com/environment/2019/oct/09/revealed-20-firms-third-carbon-emissions

33   The Carbon Majors Database: CDP Carbon Majors Report 2017, https://b8f65cb373b1b7b15feb-c70d8ead6ced550b4d987d7c03fcdd1d.ssl.cf3.rackcdn.com/cms/reports/documents/000/002/327/original/Carbon-Majors-Report-2017.pdf

34   '101 Companies Committed To Reducing Their Carbon Footprint', *Forbes*, August 2019, https://www.forbes.com/sites/blakemorgan/2019/08/26/101-companies-committed-to-reducing-their-carbon-footprint/#3b3e99f3260b

act. It was people — and especially tangata whenua — who secured the ban on offshore oil and gas exploration, and it was young Kiwis who came up with the idea of the Zero Carbon Act, and who pushed until 119 MPs voted for it.

'Ultimately, even the Paris Agreement itself is proof of the power people have in pushing decision-makers,' Tong says. 'Perhaps the most important thing about individual action is that it makes us more compelling advocates for systemic change. Research shows that people trust calls for climate action and justice more when the person making the call is walking the talk.'

So, how can people effect change?

Aside from voting carefully and joining protests when they happen, Tong recommends using social media to apply direct pressure, emailing and phoning local MPs and district councillors, and getting involved with advocacy volunteer groups.

'And volunteering for a group like a non-government organisation doesn't have to come with a scary, psychological barrier,' he says. 'These groups generally just believe in good things and want to make them happen.'

It's worth, too, noting another psychological barrier we all need to get past.

Scientists rightly call climate change a 'wicked problem', as science itself can't overcome it — especially when we begin normalising climate conditions we shouldn't be normalising. In one fascinating study, a group of US researchers illustrated this danger by quantifying a timeless and universal pastime — talking about the weather — using an analysis of posts on Twitter.

They sampled 2.18 billion geolocated tweets created between March 2014 and November 2016, to determine what kind of temperatures generated the most posts about weather.[35] They found

---

35   'Tweets tell scientists how quickly we normalize unusual weather', *Science Daily*, February 2019, https://www.sciencedaily.com/releases/2019/02/190225170252.htm

that people often tweet when temperatures are unusual for a particular place and time of year — a particularly warm March or unexpectedly freezing winter, for example. However, if the same weather persisted year after year, it generated less comment on Twitter, indicating that people began to view it as normal in a relatively short amount of time.

This phenomenon, the authors noted, was a classic case of the boiling frog metaphor. If a frog jumps into a pot of boiling-hot water, it immediately hops out. If, however, the water in the pot is slowly warmed to a boiling temperature while the frog is in it, it doesn't hop out, and is eventually cooked.

A similar problem is climate apathy. Journalists may have largely learned to ignore the misguided ramblings of cranks who reject climate science, but the apocalyptic narratives that often colour our reporting can only deepen that public sense of hopelessness. Because the potentially devastating consequences of global warming threaten our fundamental tendency to see the world as safe, stable and fair, people often respond by discounting the evidence, or by simply saying that the problem is too big. A case in point is the poll mentioned above, indicating that most Kiwis feel we won't be able to avert catastrophe.

Research suggests that if the media — and scientists — avoid doomsday narratives and focus on positive messages, people will not only be more receptive of the evidence, but will also be more willing to reduce their carbon footprint.

Kiwis now overwhelmingly accept that evidence, and recognise the threat.

And we can indeed say this: Kiwis now overwhelmingly accept that evidence, and recognise the threat. But, as Tong notes, while awareness can drive action, it doesn't automatically lead to action.

'People change their behaviour when they see a problem, and see how they can be part of fixing it,' he says. 'It's not enough to show people that we face a climate crisis. We also need to build a new narrative of how we can solve this crisis together.'

Climate, agriculture and food: How to turn the triple threat into a three-fold solution

ROD ORAM
Business journalist, *Newsroom*

C limate, agriculture and food have a deeply symbiotic relationship. Yet, for much of human history we have usurped nature. The way we produce food often degrades land and water, depletes ecosystems, reduces biodiversity and contributes to the climate crisis. We have spectacularly increased the volume and affordability of food. But the harder we've pushed food's biological and economic systems, the more unsustainable they've become. This vicious cycle is exacerbating the damage we are causing to nature and climate, and to farming and food production.

Given that severe climate change will be irreversible, humankind has only a few decades left to massively transform agriculture and food production so that it works with nature and not against it. In doing so, we will help ecosystems restore themselves so that they can help us solve our climate crisis. By nurturing nature in this virtuous cycle, we will also have enough nutritious, sustainable food to feed 10 billion people by 2050 — a 30 per cent increase from today's population.

Here in Aotearoa, we have a distinctive role to contribute to this worldwide regeneration of agriculture and food. Ours was the last large landmass to be settled by humans, less than a millennium ago. Our ancestors found it rich in abundant and diverse species, many of them unique to our land, rivers and seas. Yet, within a few centuries we had dramatically changed and depleted ecosystems. Ours has been one of the fastest degradations of nature in human history.

Remnants of those once bountiful ecosystems, home now to mostly threatened indigenous species, are all we have left. Yet, they are just enough to give nature a chance to recover if we work wisely with it. We can transform our land use, agriculture and food production so that they benefit our indigenous species and ecosystems. We can turn one cause of the climate crisis into a solution.

Coupled with New Zealand's proven record of innovation in agriculture and species protection, this is a compelling competitive advantage for our food producers and a rich source of identity for us as a nation. In contrast, new food technologies such as growing plants in climate-controlled buildings or growing animal and fish flesh from stem cells can at best only reach zero environmental impact. They cannot actively contribute, as our renewed farming systems can, to ecosystem restoration to aid climate mitigation and adaptation.

The enormity of our interdependent agricultural, food and climate challenges has been incontrovertibly documented by scientists in myriad reports in recent decades. For example, the Fifth Assessment Report (AR5) by the United Nations' Intergovernmental Panel on Climate Change (IPCC), finalised in 2014, dealt extensively with these issues. AR5 helped inform the UN's climate negotiations in Paris in 2015, at which nations committed to reducing their greenhouse gas emissions.

In 2019, the IPCC's Special Report on Climate Change and Land[1] produced further analysis which showed that the climate crisis is damaging the ability of the land to sustain humanity, with cascading risks becoming increasingly severe as global temperatures rise. Continuing destruction of forests and huge emissions from cattle and other intensive farming practices will intensify the climate crisis, making the impacts on land still worse, it concluded.

In aggregate, farming and related land-use practices account for almost one-quarter of human-related greenhouse gases. Dr Rattan Lal, the 2020 World Food Prize Laureate[2] and a professor of soil science at Ohio State University, has calculated that over the past 150 years, 476 gigatonnes[3] of carbon have been emitted from farmland soils due to inappropriate farming and grazing practices, compared with 'only' 270 gigatonnes emitted from burning fossil fuels.

---

1   This can be viewed at https://www.ipcc.ch/srccl/
2   'Soil Prof Hits Pay Dirt: $250K Prize For Helping Farmers, Fighting Climate
     Change', National Public Radio, 22 June, 2020. https://www.npr.org/sections/
     goatsandsoda/2020/06/22/880932230/soil-prof-hits-pay-dirt-250k-prize-for-helping-
     farmers-fighting-climate-change
3   1 gigatonne is equal to 1,000,000,000 tonnes.

FACTS ABOUT FOOD AND CLIMATE CHANGE

The United Nations' Food and Agriculture Organization (FAO) presented the top 10 facts on food and climate change to the 2015 Paris climate negotiations.[4] They were:

- 75 per cent of the world's poor and food-insecure people rely on agriculture and natural resources for their livelihoods.
- The FAO estimates that world food production must rise 60 per cent to keep pace with demographic change. Climate change puts this at risk.
- According to the IPCC, crop-yield declines of 10–25 per cent may be widespread by 2050, due to climate change.
- Rising temperatures are predicted to reduce catches of the world's main fish species by 40 per cent.
- Although global emissions from deforestation have dropped, deforestation and forest degradation still account for 10–11 per cent of global greenhouse gas emissions. Emissions from forest degradation (logging and fires) increased from 0.4 to 1.0 gigatonnes of carbon dioxide per year between 1990 and 2015.
- Livestock contributes nearly two-thirds of agriculture's greenhouse gases and 78 per cent of its methane emissions. Livestock grazing occupies 70 per cent of all land used for agriculture, or 30 per cent of the land surface of the Earth.
- Climate change can transfer risks of food-borne diseases from one region to another, threatening public health in new ways.
- The FAO estimates that the potential to reduce greenhouse gas emissions from livestock production (methane especially) is about 30 per cent of baseline emissions.
- Currently, one-third of the food we produce is either lost or wasted. The global cost of food wastage is about US$2.6 trillion per year, including US$700 billion of environmental costs and US$900 billion of social costs.
- Global food loss and waste generate about 8 per cent of humankind's annual greenhouse gas emissions.

---

4   'Climate Change and Your Food, Ten Facts,' UN Food and Agriculture Organisation, http://www.fao.org/news/story/en/item/356770/icode/

Moreover, greenhouse gas emissions from manufactured phosphorus and nitrogen fertilisers have risen ninefold since the early 1960s. Their flows through ecosystems constitute the largest of the two breaches of the nine planetary environmental boundaries so far catalogued by Earth system scientists led by the Stockholm Resilience Centre. The other breach is the loss of genetic diversity because of the extinction of species, while land-use changes are close to becoming the third breach. Farming practices are the dominant driver of these three degradations.

While global agricultural output has grown by between two and a half and three times in the past 50 years,[5] keeping pace with the growth in human population, some of that food consists of 'empty calories', those which provide energy but little or no other nutrition. Consequently, there are now more obese people (due to a number of causes, not just nutrition) in the world than malnourished people. This is causing a health crisis.

'Overweight and obesity are linked to more deaths worldwide than underweight. Globally there are more people who are obese than underweight — this occurs in every region except parts of sub-Saharan Africa and Asia,' the World Health Organization reported in April 2020.[6]

The interlinked crises of climate and food are the subject of an increasing number of studies. The most comprehensive and influential to date is the EAT-Lancet Commission, a joint project of EAT, a Scandinavian NGO focused entirely on food-system reform, and *The Lancet*, the British peer-reviewed medical journal. It involved 37 leading scientists across relevant disciplines from 16 countries, and it delivered its report in early 2019.[7]

---

5   *Sustainable Food Production: Fact and Figures*, SciDevNet. https://www.scidev.net/global/food-security/feature/sustainable-food-production-facts-and-figures.html
6   *Obesity and overweight*, World Health Organisation, April 2020. https://www.who.int/news-room/fact-sheets/detail/obesity-and-overweight
7   *The Lancet*'s version for scientific readers https://www.thelancet.com/commissions/EAT ; and the EAT's version for general readers https://eatforum.org/eat-lancet-commission/

FUNDAMENTAL FACTORS DRIVING THE INTERLINKED FOOD AND CLIMATE CRISES

The EAT-Lancet Commission identified nine fundamental factors driving the inter-linked food and climate crises:

- In the past 50 years, global food production and dietary patterns have changed substantially. Focus on increasing crop yields and improving production practices has contributed to reductions in hunger, improved life expectancy, falling infant and child mortality rates, and decreased global poverty.
- However, these health benefits are being offset by global shifts to unhealthy diets that are high in calories and heavily processed and animal-sourced foods. These trends are driven partly by rapid urbanisation, increasing incomes, and inadequate accessibility of nutritious foods.
- Transitions to unhealthy diets are not only increasing the burden of obesity and diet-related non-communicable diseases but are also contributing to environmental degradation.
- The human cost of our faulty food systems is that almost 1 billion people are hungry, and almost 2 billion people are eating too much of the wrong food. The Global Burden of Disease Study indicates dietary factors as a major contributor to levels of malnutrition and obesity . . . the burden of non-communicable diseases is increasing, and unhealthy diets account for up to 11 million avoidable premature deaths per year.
- Food production is the largest cause of global environmental change. Agriculture occupies about 40 per cent of global land, and food production is responsible for up to 30 per cent of global greenhouse gas emissions and 70 per cent of freshwater use.
- Conversion of natural ecosystems to croplands and pastures is the largest factor causing species to be threatened with extinction. Overuse and misuse of nitrogen and phosphorus causes eutrophication and dead zones in lakes and coastal zones.

- Environmental burden from food production also includes marine systems. About 60 per cent of world fish stocks are fully fished, more than 30 per cent overfished, and catch by global marine fisheries has been declining since 1996.
- Agricultural production is at the highest level it has ever been, but is neither resilient nor sustainable, and intensive meat production is on an unstoppable trajectory comprising the single greatest contributor to climate change. Industry too has lost its way, with commercial and political interests having far too much influence, with human health and our planet suffering the consequences.
- Transformation to healthy diets by 2050 will require substantial dietary shifts, including a greater than 50 per cent reduction in global consumption of unhealthy foods, such as red meat and sugar, and a greater than 100 per cent increase in consumption of healthy foods, such as nuts, fruits, vegetables and legumes.

The Lancet's editorial accompanying the study's release declared: 'Civilisation is in crisis. We can no longer feed our population a healthy diet while balancing planetary resources. For the first time in 200,000 years of human history, we are severely out of synchronisation with the planet and nature. This crisis is accelerating, stretching Earth to its limits, and threatening human and other species' sustained existence.'

The report's data, analysis and graphics are compelling. For example, red-meat consumption in North America is five times the recommended healthy intake per person; in Europe and central Asia it is three times. Dairy consumption in those three regions is only moderately above the recommended intake, but the adverse environmental impacts of dairy production per serving are similar to red meat, so they share the same transformational challenge.

The report concluded that sustainable food production for about 10 billion people should use no additional land, safeguard existing biodiversity, reduce water consumption and manage water responsibly, substantially reduce nitrogen and phosphorus pollution, produce zero carbon dioxide emissions, and cause no further increase in methane and nitrous oxide emissions.

It noted: 'Transformation to sustainable food production by 2050 will require at least a 75 per cent reduction of yield gaps, global redistribution of nitrogen and phosphorus fertiliser use, recycling of phosphorus, radical improvements in efficiency of fertiliser and water use, rapid implementation of agricultural mitigation options to reduce greenhouse-gas emissions, adoption of land management practices that shift agriculture from a carbon source to sink, and a fundamental shift in production priorities.'

The planetary health diet advocated by the EAT-Lancet project is a global reference diet for adults that is symbolically represented by half a plate of fruits, vegetables and nuts. The other half consists of primarily whole grains, plant proteins (beans, lentils, pulses), unsaturated plant oils, modest amounts of meat and dairy, and some added sugars and starchy vegetables. The diet is quite flexible and allows for adaptation to dietary needs, personal preferences and cultural traditions. Vegetarian and vegan diets are two healthy options within the planetary health diet but are personal choices.

Some scientists have challenged the project on various aspects of its analysis, such as on the food-related burden of disease;[8] of advocating a reduction in meat and dairy consumption when those foods offer high-density nutrition, particularly of essential micronutrients; of low-income people facing higher food costs compared with their existing diets; and for the practicality of changing farming practices to achieve sufficient supply. The projects' corresponding authors have responded in various

---

8   Burden of disease is a measure of population health that aims to quantify the gap between the
       ideal of living to old age in good health, and the current situation where healthy life is shortened
       by illness, injury, disability and premature death.

forums, such as articles in *The Lancet* on their dietary assessments[9] and on their burden-of-disease estimates.[10]

We can't achieve the linked goals of healthy people and a healthy planet by incremental improvements to existing systems. They are too broken, their damage is too great and our time too limited. A wealth of investigations, initiatives and organisations have embraced this essential truth in recent years. Examples include the work of the World Economic Forum,[11] the Commonwealth's regenerative agricultural programme for small-scale farmers in member countries,[12] the Food and Land Use Coalition,[13] and major multinational food companies such as Unilever and Danone, which are both heavily funding regenerative agricultural strategies.[14, 15]

The term 'regenerative agriculture' (regen ag) was conceived by the late Bob Rodale in the 1970s. Growing up on a farm in Pennsylvania, he became a leader, through his work as an advocate and publisher, on agricultural practices that generate greater health and wellbeing for people and ecosystems. Within the term, he summarised a set of farming principles and practices that enrich soils, improve watersheds, enhance ecosystem services such as soil carbon and nitrogen sequestration, improve biodiversity, and promote farmer and livestock welfare. Thirty years on from Rodale's death, the institute

9   'Healthy diets and sustainable food systems — Authors' reply', *The Lancet*, 21 June 2019.
        https://www.thelancet.com/journals/lancet/article/PIIS0140-6736(19)31101-8/fulltext
10   'The EAT–Lancet Commission: a flawed approach? — Authors' reply', *The Lancet*, 28 September
        2019. https://www.thelancet.com/journals/lancet/article/PIIS0140-6736(19)31910-5/fulltext
11   *The Food Action Alliance*, World Economic Forum. https://www.weforum.org/
        press/2020/01/global-leaders-unite-under-the-food-action-alliance-to-deliver-a-better-
        future-for-the-people-and-the-planet/
12   Common Earth: https://thecommonwealth.org/media/news/game-commonwealth-steps-
        battle-climate-change-regenerative-solutions-model
13   *Growing Better: Ten Critical Transitions to Transform Food and Land Use*, The Food and Land
        Use Coalition. https://www.foodandlandusecoalition.org/global-report/
14   Unilever press release, 15 June 2020. https://www.unilever.com/news/press-
        releases/2020/unilever-sets-out-new-actions-to-fight-climate-change-and-protect-and-
        regenerate-nature-to-preserve-resources-for-future-generations.html
15   *Regenerative agriculture*, Danone. https://www.danone.com/impact/planet/regenerative-
        agriculture.html

that bears his name remains a leader in this fast-expanding field.[16]

Regenerative practices have much in common with organic, biodynamic and other systems that work in natural ways with the complexity of ecosystems. Regen ag severely minimises or eliminates artificial fertilisers, agrichemicals and other manufactured compounds, although practices vary according to climate, terrain, the crops and animals raised and many other factors.

In terms of our response in Aotearoa New Zealand to these intense global challenges and abundant opportunities, the best place to start is with the state of our environment. Its health is fundamental to the four capitals — natural, financial/physical, social and human — on which our wellbeing depends. These in turn form the basis of our government's Living Standards Framework[17] and related Wellbeing Budget process.[18] Moreover, we strongly identify ourselves as a nation by our wild and rural places, even though a higher proportion of us live in towns and cities compared with many other developed countries.

The most comprehensive single source of environmental information and analysis is 'Environment Aotearoa', a synthesis report produced every three years by the Ministry for the Environment and Statistics New Zealand. It brings together the agencies' periodic domain reports on the marine environment, freshwater, atmosphere and climate, land, and air. The 2019 edition[19] concluded that many measures of environmental health continued to decline. Some of these factors also contributed to climate change, which in turn exacerbated the overall decline of the environment.

The MFE has also produced a broad summary[20] of the predicted

---

16   https://rodaleinstitute.org/

17   Living Standards Framework – Dashboard, The Treasury. https://lsfdashboard.treasury.govt.nz/
        wellbeing/

18   Wellbeing Budget 2020, The Treasury. https://www.treasury.govt.nz/publications/wellbeing-
        budget/wellbeing-budget-2020

19   'Environment Aotearoa 2019', Ministry for the Environment, April 2019. https://www.mfe.govt.
        nz/environment-aotearoa-2019-summary

20   'Likely climate change impacts in New Zealand', Ministry for the Environment. https://www.mfe.
        govt.nz/climate-change/likely-impacts-of-climate-change/likely-climate-change-impacts-nz

impacts of climate change out to 2090. These will vary across the country, as will their influence on agriculture. Temperatures will rise, with greater increases in the North Island than the South, and the greatest warming in the northeast. But the amount of warming in New Zealand is likely to be lower than the global average.

Sea levels will rise. Extreme weather events will become more frequent, both droughts (especially in the east of New Zealand) and floods. Changes in rainfall patterns will bring increased summer rainfall in the north and east of the North Island, and increased winter rainfall in many parts of the South Island.

Warmer temperatures will alter habitats that are critical to some species, increasing the risk of localised extinction. Warmer temperatures will favour conditions for many exotic species. They will also favour conditions for the spread of disease and pests affecting both fauna and flora.

The climate impacts on biodiversity will be significant. Warmer temperatures will alter habitats that are critical to some species, increasing the risk of localised extinction. Warmer temperatures will favour conditions for many exotic species. They will also favour conditions for the spread of disease and pests affecting both fauna and flora. Increased summer drought will put stress onto dry lowland forests. Earlier springs and longer frost-free seasons could affect the timing of bird egg-laying, first flowering and the health of leafing or flowering plants.

Agricultural productivity is expected to increase in some areas. However, there are the aforementioned risks of drought and the spreading of pests and diseases. There are likely to be costs associated with changing land-use activities to suit a new climate.

Urban New Zealand faces many of the same challenges of adapting to climate change, such as the impact of rising sea levels and more-extreme weather events, particularly on its built environments.

It also has some challenges such as decarbonising transport, industrial processes and high emissions steel and concrete.

In terms of mitigation, the current national goals in our Zero Carbon Act 2019 are to reduce carbon dioxide emissions to net zero by 2050. This excludes biogenic methane from ruminant animals and from landfill and other sources. The target for them is a 24–47 per cent cut from 2017 levels by 2050. However, our only ratified, unconditional commitment under the Paris Agreement is to a 30 per cent reduction in emissions from 2005 levels by 2030.

The task of reducing greenhouse gas emissions is split almost fifty-fifty between urban and rural sources. In total our gross emissions increased 20.4 per cent from 1990 to 2018, to 15.3 million metric tonnes of carbon dioxide equivalents (Mt CO^2e). This increase was mostly due to increases in methane from dairy cattle digestive systems and carbon dioxide from road transport.

Of the agricultural component, methane from our ruminant animals accounted for 36.5 per cent of the nation's total emissions in 2018, nitrous oxide from urine and fertiliser 6.3 per cent, and greenhouse gas emissions from other agriculture sources 4.9 per cent.

While methane is a short-lived greenhouse gas compared with carbon dioxide, it is a far more powerful driver of climate change. As Simon Upton, the Parliamentary Commissioner for the Environment, wrote in a 2018 report: 'A constant flow of methane emissions results in a constant methane concentration after around fifty years, but its impact on temperature continues to increase for several centuries. Three hundred years after a constant flow of methane emission has started, the warming effect is more than twice as high as it is after fifty years.'[21]

Thus, '[I]f New Zealand's emissions of livestock methane were

21   'A note on New Zealand's methane emissions from livestock', Parliamentary Commission for the Environment, p. 8.
        https://www.pce.parliament.nz/publications/a-note-on-new-zealand-s-methane-emissions-from-livestock

held steady at 2016 levels, then within about ten years the amount of methane in the atmosphere from that source would level off. However, the warming effect of that methane would continue to increase, at a gradually declining rate, for more than a century. In the year 2050, holding New Zealand's livestock methane steady at 2016 levels would cause additional warming of 10–20 per cent above current levels. This warming would increase to 25–40 per cent by 2100.'[22]

New Zealand's contribution to global warming is disproportionately larger than our share of the world's population or land area, the New Zealand Agricultural Greenhouse Gas Research Centre noted in July 2019: 'While small in absolute terms, New Zealand's share in global warming to date is more than 4 times greater than its share of the global population and about 1.5 times greater than its share of the global land area.'[23]

It also highlighted methane's prominent role in our climate change profile. 'New Zealand's biogenic methane emissions currently make a bigger estimated contribution to global warming than cumulative [New Zealand] emissions of fossil carbon dioxide and nitrous oxide combined. If gross emissions of those three gases continued at current rates, biogenic methane would remain New Zealand's largest single contributor to global warming for the next six decades despite its relatively short lifetime in the atmosphere compared to carbon dioxide and nitrous oxide.'

Given the enormity of such issues, devising fair and effective strategies for tackling climate change is proving a fraught political, economic and social exercise in all nations. New Zealand is no exception. It has taken us the first two decades of this century to achieve only broad support for some long-term goals for emission reductions, a Climate Change Commission to help guide us, and an effective Emissions

---

22   Ibid., p. 11.
23   NZ Agricultural Greenhouse Gas Research Centre, October 2019. https://www.nzagrc.org.nz/knowledge,listing,593,scientific-aspects-of-new-zealands-2050-emission-targets.html

Trading Scheme to incentivise us to change.

That leaves us barely a decade to make our first deep cuts in emissions, so we then have some chance of reaching our goal of net zero emissions, excluding biogenic methane, by 2050. And even if all nations achieve that target, we humans cannot be sure we can keep the rise in global temperatures to 1.5°C. Above that threshold, changes in climate and ecosystems that are deeply damaging to all life forms on the planet will accelerate greatly.

The sheer speed, scale and complexity of the change we have to achieve in values, behaviour, technology and economics is daunting. As the New Zealand Productivity Commission wrote in its August 2017 issues paper on our transition to a low-emissions economy: '[T]he shift from the old economy to a new, low-emissions economy will be profound and widespread, transforming land use, the energy system, production methods and technology, regulatory frameworks and institutions, and business and political culture.'[24]

Yet, despite the urgent need for such transformational change, all we've implemented so far are overall frameworks and mechanisms such as the Climate Commission and the ETS. We have yet to put in place programmes, regulations, incentives and other mechanisms which will actually cut emissions. Many other countries are similarly underperforming.

In the case of our agricultural sector, there is widespread agreement that the climate is changing. But our dairy and red-meat producers have been offering only modest improvements, not systemic reform, arguing that they have two advantages over their competitors abroad:

---

24   *Lower Emissions Economy — Issues Paper*, New Zealand Productivity Commission, August 2017, p. 1. https://www.productivity.govt.nz/assets/Documents/50449807ff/Low-emissions-economy-issues-paper.pdf

- They are more efficient, and their pasture-based systems have lower adverse environmental impacts compared with those used by farmers overseas, particularly feedlots and other very high-intensity operations. Thus, they argue that any reduction in farming emissions in New Zealand will cause production to fall here and to rise on farms overseas, leaving the world worse off.
- They believe they will always have plenty of consumers overseas who are willing to pay high prices for their high-quality products. New Zealand's 'clean, green' image, and their farms' relatively lower emissions profile will only enhance their marketability.

This complacency is unfounded:

- Many consumers here and abroad are becoming more conscious about their food, particularly in terms of its nutritional, animal welfare and environmental attributes; as a consequence, their diets are becoming more diverse.
- We produce a minute proportion of world food supplies. Even our dairy industry's share of global output is only 2 per cent, equal to just one year's increase in global supply. Thus, our biggest contribution globally is to pioneer new, climate-compatible farming systems, not to use the current problematic practices to maintain our carbon-intensive exports.
- Our red-meat and dairy farmers have long used existing technologies and practices to improve their environmental performance. Both sectors can make further progress by widespread adoption of these best practices while scientists develop greater advances.
- The increasing emphasis on nature-based responses to the climate crisis, such as increasing carbon sequestration by soils, plants and forests, complements essential technological shifts, such as from

fossil fuels to clean energy. Thus farmers, particularly in the dairy and red-meat sectors, have a significant role to play. By reducing their farms' emissions and increasing their carbon sequestration, they would begin to turn farming from an extractive and polluting system into a regenerative one, which would be more productive and environmentally sustainable. For example, sequestering more carbon in soils and plants increases the fertility of soils and the biodiversity of ecosystems.

- Plant-based and stem-cell-grown products are increasingly similar in look, taste and texture to meat and dairy products from ruminant animals, but with significantly less adverse environmental impacts. The plant ones are already cost-competitive and widely available in supermarkets here and abroad; the ones from stem cells will be in coming years.

- The agricultural sector accounts for only a small share of our total carbon dioxide emissions. But it too can sharply cut its emissions while being rewarded for sequestering more carbon dioxide once ways are developed, for example, of measuring carbon in soils and small strips of forestry.

Moreover, our farmers have long made progress on these issues. For the past 25 years they have averaged a 1 per cent a year cut in methane emitted per litre of milk and kilogram of red meat — and have enhanced their farm profitability in the process; the best farmers are achieving faster reductions. One example is Owl Farm near Cambridge, which is one of the commercial farms monitored by DairyNZ, the sector's farmer-funded research and advocacy arm.

The Interim Climate Change Committee (the precursor to the Climate Change Commission, established by our zero-carbon legislation in 2020) reported in 2019 that Owl Farm had reduced total greenhouse gas emissions by 8 per cent, while lifting operating profit per hectare by 14 per cent through improved management practices. '[F]urther farm management changes involving reduced feed use and lowering the stocking rate is expected to increase profitability by another 21 per cent, [and] reduce nitrate leaching by 14 per cent and greenhouse gas emissions by 13 per cent.' Owl Farm noted there was 'a potential [positive] downstream economic impact of reducing the intensity of their farming operation'.[25]

A farm's ability to increase its profitability through improving its environmental performance by in part reducing the intensity of its stocking and feeding was a dynamic proven in a 2004 report by the then Parliamentary Commissioner for the Environment, Morgan Williams.[26] Over the nearly two decades since, some dairy farmers have de-intensified their operations. But intensification, with adverse environmental and economic consequences, particularly high-debt levels on some farms, has been the sector's dominant business model.

Owl Farm's experience was also consistent with a 2018 report of the Biological Emissions Reference Group (BERG).[27] 'If all farmers operated using today's best practice, we may be able to reduce emissions by up to 10 per cent,' the Ministry for Primary Industries said when releasing the report.[28]

Looking further ahead, BERG expressed 'medium to high confidence' that a vaccine with the potential to deliver a 30 per cent reduction in biogenic methane would be available by 2050, and 'high confidence' that a methane inhibitor for grazing systems could do the same by 2050.

---

25   *Action on Agricultural Emissions,* Interim Climate Change Committee, April 2019, page 3. https://www.iccc.mfe.govt.nz/assets/PDF_Library/f15921453c/FINAL-ICCC-Agriculture-Report.pdf
26   *Growing for Good: Intensive Farming, sustainability and New Zealand's Environment,* Morgan Williams, October 2004.
27   Report of the Biological Emissions Reference Group, December 2018. https://www.mpi.govt.nz/dmsdocument/32125-berg-report-final-for-release-6-dec
28   *Government and industry partners release report on biological emissions,* Ministry for Primary Industries, 6 December 2018. https://www.mpi.govt.nz/news-and-resources/media-releases/government-and-industry-partners-release-report-on-biological-emissions/

The report showed that farmers could easily meet what subsequently became the country's 2030 goal of cutting methane by 10 per cent from 2017 levels by simply adopting *current* best farming practice, as Owl Farm and others demonstrate, and they had a good chance of *exceeding* New Zealand's current goal of reducing methane by between 24 and 47 per cent by 2050 if the new technologies it identified lived up to their promise.

BERG was a two-year, joint agriculture industry–government working group of nine key players: Beef+Lamb, DairyNZ, Deer Industry New Zealand, Federated Farmers, the Fertiliser Association of New Zealand, Fonterra, Horticulture New Zealand, and the ministries for Primary Industries and the Environment, which all signed off on the report. But within six months, most of those industry organisations were strongly opposing the government's proposed methane targets.

Federated Farmers leads the sector's most aggressive opposition to on-farm action on climate change. It says its members fully support New Zealand's efforts to meet our international climate commitments — but it says the 'limited tools in the toolbox' means that the maximum

Federated Farmers leads the sector's most aggressive opposition to on-farm action on climate change.

our farmers can achieve is a 3 per cent reduction of methane by 2030, and 10 per cent by 2050.

Worse, it denies the actual progress long under way in our agricultural science and farming practices. In its submission on the 2019 Zero Carbon Bill, Federated Farmers said the results of an online survey of its members showed that a high proportion of them were strongly opposed to the targets for methane reductions.

While Federated Farmers seems to attract the more militant farmers, the wider sector appears to have a more positive, though still cautious, engagement on climate issues, judging by a 2019 survey conducted for the Ministry for Primary Industries.[29] This found that 63 per cent of farmers agree or strongly agree that global human activity was contributing to climate change, up from 54 per cent in the previous survey, in 2009. But the 2019 figure was lower than the 82 per cent of *all* New Zealanders who believe that human activity was at least partly contributing to climate change.

About half the farmers surveyed felt that their farms were already being affected by climate change and severe weather patterns, and thought climatic changes presented both an opportunity and a threat to their business. Yet a similar proportion felt that the agricultural sector was doing enough to adapt to the environmental impacts of more-severe weather patterns and changing climatic conditions. The survey also showed that many farmers had minimal knowledge of their farms' emissions and a limited understanding of how to reduce them.

In contrast, there is a small but growing number of eager adopters of emission-reduction practices. They range from small operations such as Owl Farm (mentioned above) to Landcorp, the state-owned enterprise which is the largest farmer in the country.

In 2015, Landcorp set itself the goal of being a carbon-neutral farmer by 2025. To that end, it is, for example, reducing forages dependent on high levels of nitrogen fertiliser (irrigated ryegrass, for example, needs on average 250 kg of nitrogen per ha per year, whereas lucerne requires none); using livestock genetics to help it breed animals with inherently lower emissions than average to repopulate its herds; beginning the shift to regenerative agricultural practices such as reducing stock per hectare, low or no use of nitrogen fertilisers, no pesticides/ herbicides; building organic soil health (which sequesters carbon in the soil); diversifying plant species and integration of tree systems; and retiring marginal land into trees and other plantings, an action which

---

29   *Climate Issues Facing Farmers*, Ministry for Primary Industries, March 2019. https://www. agriculture.govt.nz/dmsdocument/33747/direct

supports biodiversity, protects water, provides shade and shelter for stock and reduces what are vulnerable areas for both stock and the environment.

Two crucial keys to the shift are 1) a reduction in the farming of dairy cows and developing instead lower-emission alternatives such as ewe and deer milk, and 2) Landcorp's use of its Pāmu brand to help it develop both a premium in the marketplace and a closer connection to consumers.

Meanwhile, across the sector there are signs of a growing engagement on farming's climate change issues. Fonterra, for example, is rolling out customised emissions reports for each of its supplying farms by the end of the 2020–21 season. However, while it acknowledges past progress — 'Over the last 25 years, New Zealand dairy farmers have reduced on-farm emissions intensity by about 20 percent. The strongest improvements were from 2007 to 2016'[30] — it is still sticking with its longstanding pledge to keep its methane levels only unchanged by 2030, not reduced.

But without strong leadership from Fonterra and its farmers, the agricultural sector won't make significant methane reductions: nor will the country meet its overall targets. The co-op and its suppliers are the largest single source of greenhouse gas emissions in New Zealand, accounting for some 20 per cent of the national total.

Yet in contrast to its suppliers' minimal progress reducing emissions on-farm, which account for 90 per cent of the co-op's emissions, Fonterra is making good progress on the other 10 per cent, which come from downstream activities such as transport, processing and packaging. It has committed to reducing those by 30 per cent from 2015 levels by 2030 and to net zero by 2050 by, for example, switching from coal to biomass fuel in some of its processing plants.

-----

30   Fonterra press release, https://www.fonterra.com/nz/en/our-stories/articles/reducing-our-emissions.html#:~:text=Over%20the%20last%2025%20years,were%20from%202007%20to%202016.

For the agricultural sector as a whole, the two biggest strategic shifts it has signalled in many years came in proposals that various farmer bodies made in 2019 and 2020. If both initiatives fulfil their potential over, say, the next decade, the sector will make substantial progress on climate change in particular, sustainability in general and branded, value-added products overall.

In the first, in mid-2019 a consortium of primary-sector organisations including DairyNZ and Beef+Lamb proposed that it and the government set up a joint body to work out how to measure, manage and reduce emissions and price them outside the Emissions Trading Scheme (ETS). The government agreed, with a 2025 deadline for completing the work. But if the government deems insufficient progress is being made by 2022, it has reserved the right to bring agriculture into the ETS.

This proposal by the sector marks its first broad support for some form of emissions pricing after two decades of fighting against it. It was prompted by the government's intention to include the sector in the ETS. The Interim Climate Change Commission had recommended the government should start charging farm emissions through the ETS from 2020 until 2025, at a heavily discounted carbon price equivalent to 1 cent on every kg of milk solids and between 1 and 4 cents per kg of meat, which would represent only a minute cost to farmers.

The body the primary sector proposed is called He Waka Eke Noa — Primary Sector Climate Action Partnership, which involves eleven sector organisations and two government departments. Its six work-streams are: farm planning; emissions reports; emission pricing; on-farm sequestration; innovation and uptake; and extension and early action. Māori agribusiness will integrate Māori perspectives into each work-stream. It began work in late 2019 and early reports suggest that it is making good progress.

The second, and more significant, shift came with the Primary Sector Council's vision and strategy document, called *Fit for a Better World,* released in July 2020. It is based strongly on the concept of regenerative agriculture, which is given a unique and deeply New Zealand context through its embrace of te taiao, an expression of Te Ao Māori, the world view of humankind's symbiotic relationship with nature. With the strategy document, the council released a companion document describing this called *Te Taiao Ora Tangata Ora — the Natural World and our People are Healthy.*

The council was sct up by the government in April 2018 with 15 members from across the sector, including chair Lain Jager, a former Zespri chief executive. Its preliminary report in December 2019 was broadly based on regenerative agriculture and te taiao. But its final document six months later explained much more fully how the two were the foundation and ethos of the sector's strategy. If regenerative agriculture and te taiao flourish across the sector, they would create a powerful and unique brand for New Zealand food in world markets. Likewise, the endorsement of these principles by the Ministry for Primary Industries and other government agencies was more comprehensively described. They have begun work on creating practical programmes to turn the vision and strategy into reality.

Over recent years, awareness of te ao Māori has been embraced by more and more organisations, such as the Reserve Bank and the 11 National Science Challenges.

They see wisdom in the Māori understanding that we humans are inherently a dependent part of all living systems.

Even more importantly, these and other institutions are taking a great interest in te ao Māori not just to help Māori strengthen their culture and communities — they are also starting to assimilate it to improve themselves overall. They see wisdom in the Māori understanding that we humans are inherently a dependent part of all living systems. We are not the dominant force with free rein to exploit the natural world, without limits.

In contrast, the Western world view of dominion over nature was certainly a factor helping to drive humankind's vast expansion of knowledge, technology, economies, resource extraction and pollution over the past few centuries. But there is absolutely no doubt in science that we have hit the limits of such exploitation. As described earlier in this chapter, one of the best of many guides to this is the planetary boundaries research of the Stockholm Resilience Centre.

Aotearoa, too, is breaching some of these boundaries. 'New Zealand's growth model . . . has started to show its environmental limits, with increased greenhouse gas emissions, freshwater contamination and threats to biodiversity,' the OECD wrote in 2017 in its once-a-decade environmental report.[31]

'Addressing GHG [greenhouse gas] emissions from agriculture, and especially dairy farming, should remain a priority . . . [and] the need to further explore the economic opportunities that more sustainable uses could yield.'

Above all: 'Developing a long-term vision for a transition towards a low-carbon, greener economy would help New Zealand defend the "green" reputation it has acquired at an international level.'

But our understanding of these challenges, and the language to express it, has yet to advance sufficiently, either here or abroad. In many people's values and practices, stewardship of nature has displaced dominion over it. Yet that still leaves them with the hubris of believing that humans can control ecosystems. In fact, living systems are so vastly complicated that we have to learn how to watch, listen, learn and respond to them. Then, as they recover, so will we.

Our oneness with nature is at the heart of te ao Māori, as a helpful series of articles in *Te Ara*, the online encyclopaedia, explains.[32] In turn,

---

31   OECD Environmental Performance Reviews: New Zealand 2017.
      https://www.oecd.org/newzealand/oecd-environmental-performance-reviews-new-
      zealand-2017-9789264268203-en.htm
32   *Te Ao Mārama, Te Ara.* https://teara.govt.nz/en/te-ao-marama-the-natural-world

this world view has given rise to a great body of knowledge, mātauranga Māori, about our living systems in Aotearoa New Zealand.

Back in 1900, Āpirana Ngata, the Ngāti Porou leader and scholar, wrote of mātauranga Māori and mātauranga Pākehā and the great benefit of 'casting our nets between them', rather than fishing in one or the other. This insight is retold in a seminal lecture on mātauranga Māori given by Te Ahukaramū Charles Royal, an academic researcher and music composer, at the University of Canterbury in 2009.[33]

He noted in the lecture that: 'Mātauranga Māori today exists in a fragmentary and disorganised state; a good deal of existing mātauranga Māori concerns a world that only exists in the past and so a great deal of work needs to be done to bring about relevance and utility within this body of knowledge in our contemporary circumstances. Much has also been lost, and much mātauranga Māori has been superseded by other kinds of knowledge. So it is important not to make claims for mātauranga Māori that cannot be substantiated.'

He knew, though, that there was still plenty of extant indigenous Māori knowledge, both recoverable and yet to be learnt, which could contribute powerfully to our contemporary and future progress.

He identified three themes — searching for better relationships with the natural world, cross-boundary styles of thought and knowledge, and the revitalisation of traditional indigenous knowledge — which, when woven together, 'are the key ideas within international indigenous knowledge today'.

Royal has significantly led the growing understanding of mātauranga Māori and the contribution it can make to the western world view and science. The Key government began to integrate indigenous knowledge into science policy with its 2010 decision by Minister of Research, Science and Technology Wayne Mapp to apply

---

33   *Te Kaimānga: Towards a New Vision for Mātauranga Māori*, Te Ahukaramū Charles Royal, 2009. https://static1.squarespace.com/static/5369700de4b045a4e0c24bbc/t/5d549912 37bfcb0001a5dca3/1565825300610/CRoyal2009MacmillanBrownLecture.pdf

Vision Mātauranga across all science investment priority areas, and to establish the Vision Mātauranga Capability Fund to help further that goal. The following year, Vision Mātauranga policy was incorporated into the Crown Research Institutes' Statements of Core Purpose. This required them to enable the innovation potential of Māori knowledge, resources and people as part of their operating principles.

In 2014, the Key government created the 11 National Science Challenges: decade-long, interdisciplinary programmes to progress our most complex and important science.[34] Again, it made Vision Mātauranga a core element of each of them.

While in due course historians might consider these decisions a critical turning point in Aotearoa's cultural development, sadly for the first six or seven years these government initiatives to foster mātauranga Māori were often disparaging, box-ticking exercises. Rightly, Māori were insulted. Thankfully, though, over recent years the understanding that the two world views can inform and enhance each other has been growing for both Pākehā and Māori, whether they are in science, farming, conservation, business, government or many other fields.

Two examples from the National Science Challenges are Our Land and Water[35] and Science for Technological Innovation.[36] An example in conservation is Kotahitanga mō te Taiao,[37] an alliance of all the local government bodies and some of the iwi in the top of the South Island, and the Department of Conservation. Its focus is on landscape-scale conservation initiatives that also have environmental, social, economic and cultural benefits. Its vision is 'that our extraordinary natural heritage is flourishing, having been restored over large areas, including where people live. People live in, care for, and benefit from

---

34   https://www.mbie.govt.nz/science-and-technology/science-and-innovation/funding-information-and-opportunities/investment-funds/national-science-challenges/
35   https://ourlandandwater.nz/about-us/te-ao-maori
36   https://www.sftichallenge.govt.nz/for-researchers/vision-matauranga/
37   https://www.doc.govt.nz/contentassets/cf2bf2f877544dc29594442365ca797c/kotahitanga-mo-te-taiao-strategy.pdf

the environment in ways that bolster natural ecology together with the communities that live within them.'

Among other notable te ao Māori developments, the Reserve Bank has a strategy to help it 'to engage with the increasingly important and diverse Māori economy'.[38] One element is its history and explanation of the financial system and the Reserve Bank's role in it, which draws heavily on mātauranga Māori.[39] Similarly, Te Papa Tongarewa opened in 2019 a major permanent exhibition, *Te Taiao*,[40] which brings together mātauranga Māori and western science.

Internationally, the sustainable management of natural capital is taking on ever greater urgency, given the escalating damage our current technologies and economic systems are doing to ecosystems. The Natural Capital Coalition[41] is a leading example of major businesses committed to pioneering solutions.

The World Bank estimates that New Zealand ranks eighth out of 120 countries in natural capital per person, outranked only by petroleum-exporting countries. Yet we know we are rapidly degrading and depleting that rich gift.

The World Bank estimates that New Zealand ranks eighth out of 120 countries in natural capital per person, outranked only by petroleum-exporting countries.

In response, a coalition of business and some government leaders established The Aotearoa Circle in October 2018. An example of its work was the Fenwick Forum, named after the late Sir Rob Fenwick, who co-founded the circle with British sustainability leader

---

38   *Te Ao Māori: an evolving and responsible strategy*, Reserve Bank of New Zealand. https://www.rbnz.govt.nz/about-us/te-ao-maori-strategy

39   *The Journey of Te Pūtea Matua: our Tāne Mahuta*, Reserve Bank of New Zealand. https://www.rbnz.govt.nz/about-us/the-journey-of-te-putea-matua-our-tane-mahuta

40   *Te Taiao I Nature*, Te Papa Tongarewa. https://www.tepapa.govt.nz/visit/exhibitions/te-taiao-nature

41   https://naturalcapitalcoalition.org/coalition-organizations/

Sir Jonathon Porritt. It led a series of online meetings during the first Covid-19 lockdown in June 2020, to identify actions New Zealand could take to create a more productive, sustainable and inclusive economy.

The final report[42] focused on how to transform our food, transport and energy systems, and on potential partners who could collaborate to do so. Importantly, the ideas of young leaders involved in the programme were well integrated into the report. Regenerative agricultural principles were at the heart of the food-system strategy forum participants devised.

Over the past few years, an increasing number of New Zealand's government agencies, research institutes and farm advisers have started programmes on regenerative agriculture. For example, the Our Land and Water National Science Challenge in early 2020 formed a research partnership[43] with the NEXT Foundation, with funding from the Ministry for Primary Industries. A team led by Dr Gwen Grelet, soil ecologist at Manaaki Whenua–Landcare Research, and Samuel Lang, manager of Quorum Sense, a network of farmers developing regenerative practices, will develop a framework for collecting scientific evidence on the effectiveness of those practices.

'There is a growing body of anecdotal evidence for the multiple benefits of regen ag, be it environmental, economic, social, psychological or cultural, but scientific studies are scarce and restricted to overseas systems. So we don't have enough scientific data yet to compare these benefits to other current ways of farming in New Zealand,' Dr Grelet said. 'This is partly because regenerative ag emphasises adaptive management strategies that seek to optimise the performance of the whole farm, for multiple benefits simultaneously. This isn't easy to quantify using conventional academic approaches.

---

42   Aotearoa Circle Fenwick Forum, final report, June 2020. https://static1.squarespace.com/static/5bb6cb19c2ff61422a0d7b17/t/5ef16ab4eade2b25c640ad14/1592879840886/The+Fenwick+Forum+Report+June+2020.pdf
43   'What do we know about regenerative agriculture?' Our Land and Water, May 2020. https://ourlandandwater.nz/news/what-do-we-know-about-regenerative-agriculture-in-new-zealand/

'One of the most exciting things about this research is that we intend to assemble a framework that includes multiple knowledge systems, combining academic knowledge with the ways farmers know and appraise their whole system.'[44]

While only a very small proportion of Kiwi farmers are already on the journey — anecdotal evidence suggests less than 5 per cent — the number is growing rapidly. Moreover, there is a growing cohort of regen ag advisers to help them. One example is Calm The Farm,[45] a group of technical experts and investors facilitating this farming revolution. Meanwhile, Pure Advantage, a sustainable economy advocate, and the Edmund Hillary Fellowship[46] are tracking agricultural progress as part of their economy-wide project *Our Regenerative Future.*[47]

Dairy, sheep and beef farms will be the biggest beneficiaries of regenerative practices, because they have the greatest greenhouse gas problem to solve; and intensive dairying has the biggest problems to fix, of nitrogen and phosphorus polluting waterways. But arable farming and horticulture will have plenty of new opportunities, too. They will help some dairy and meat farmers diversify to reduce their adverse environmental impacts, and help to satisfy consumers' changing dietary choices.

'The opportunity for New Zealand is in manufacturing high-value plant protein foods, sourcing ingredient streams from trusted sustainable and diversified production systems that meet our future climate change challenges, and delivering premium products into the "flexitarian" diets of our international customers,' concluded scientists at Plant & Food Research, the Crown Research Institute, in their 2018 report on the horticultural potential for New Zealand.[48]

---

44   Ibid.
45   Calm the Farm. https://www.calmthefarm.nz/
46   Declaration of interest: This author is one of the EHF Fellows involved in the project.
47   *Our Regenerative Future*, https://pureadvantage.org/ourregenerativefuturecampaign/
48   Sutton, K., Larsen, N., Maggre,G.-J., Huffman, L., Clothier, B., Bourne, R., Eason, J., *Opportunities in Plant-based Foods — Protein*, Plant and Food Research, May 2018, https://www.plantandfood.co.nz/file/opportunities-plant-based-foods-protein.pdf

If most of our farmers transition to regenerative systems over the next 10 to 15 years, they will build ecological and economic resilience and establish this new competitive advantage, even over such farmers overseas. They will then have a deeply compelling story to tell to the world about their pivotal role in tackling the climate crisis; in improving the range, quality and volume of food produced here; in restoring Aotearoa New Zealand's unique ecosystems and species; and in encouraging urban Kiwis to bring true regeneration to their built environments and economic activity.

By helping nature rebuild the ecosystems on which their farming depends, our farmers will be agents of positive change. Doing so, they will build far closer relationships with their customers at home and abroad, and with their fellow Kiwis who would applaud and support such a transformation.

In all this, our symbiosis of mātauranga Māori and mātauranga Pākehā will be our powerful and distinctive contribution to knowledge and practice worldwide, as humankind seeks to rapidly turn agriculture and food production from a cause of the climate crisis to one of the solutions.

Look after nature and nature will look after us: How Aotearoa New Zealand can tackle climate change

ADELIA HALLETT
Carbon News

Here's the bottom line: nature is our best defence against the changing climate. Until we realise that, we're not going to make much progress, because it's the one solution to climate change that's guaranteed to work.

Forgetting we're part of nature is the reason we are in this mess. We're not separate from it, not immune from it and certainly not in charge of it. The world we know is a result of 4.6 billion years of interaction between the 'things' on the planet — the rocks, winds, ocean currents, rain, sun, soils, nutrients, plants and animals, from the tiniest single-cell organisms to complex creatures like us.

When you live in the modern, urban world it can be easy to forget that without these functioning Earth systems providing the food we eat, the water we drink and the air we breathe, we're toast. But that's how we've been behaving. We pave our cities and clear the hills of forests and wonder why we get floods. We strip the land and wonder why our soil fertility is declining. We allow huge amounts of nitrogen and other pollutants into our waterways and wonder why the rivers and oceans are not as clean and full of life as they used to be.

It's probably no secret to anyone now that New Zealand's natural world is in trouble. After years of selling ourselves to the world as clean and green, we're now being a lot more honest about the state of our environment. If you want details, have a look at the Environmental Reporting series by the Ministry for the Environment and Statistics New Zealand[1] or see the chapter in this book by Matt McGlone, but here are a few headlines:

1. We've already wiped out at least 75 species.
2. Around 90 per cent of native seabirds, 76 per cent of freshwater fish, 84 per cent of reptiles and 46 per cent of vascular plants are at risk of extinction.

---

1   https://www.mfe.govt.nz/more/environmental-reporting

3. Around 90 per cent of our wetlands are gone, along with more than two-thirds of our native forests.

4. Introduced grasses are now the country's most dominant plant, and nearly a tenth of the country is covered in exotic forest.

Throwing climate change into the mix takes it to another level.

The good news, though — and there *is* good news in all this — is that working with nature really will let us turn this around.

Let's start with something at the heart of life on Earth: photosynthesis, in which living organisms take carbon dioxide out of the air and convert it into the chemical energy they need for growth, expelling oxygen as a by-product. It is one of the magnificent processes that literally keeps us alive, providing us with the food we eat, the air we breathe and the comfortable climate we live in. Animals can't photosynthesise, but plants and many simple organisms like blue-green algae can.

The good news, though — and there is good news in all this — is that working with nature really will let us turn this around.

Keeping atmospheric global warming below 1.5°C, as the United Nations Framework Convention on Climate Change (UNFCCC) Special Report has so clearly told us we should,[2] requires us to not only stop sending more greenhouse gases into the atmosphere but also to actually increase the amount of carbon dioxide being taken out of it — possibly by as much as a trillion tonnes by the end of the century.[3]

And at the moment, nature is the only 'technology' proven able to do this at the scale needed. Officially, in 2018, New Zealand's forests (native and exotic) pulled nearly 17 million tonnes' worth of greenhouse gases out of the air and stored them (to put that in perspective, our total

---

2   https://www.ipcc.ch/sr15/
3   https://www.ipcc.ch/sr15/

emissions that year stood at just under 79 million tonnes).[4] The trees packed the carbon away in the cells of their branches, trunks and roots, and when they die and decay, much of the carbon will pass into the soil, making it rich and friable for other plants to grow in.

The concept of carbon forests has been around for more than a decade but is really only just starting to generate interest now that carbon prices are becoming high enough to be meaningful. I'll explain how this works in more detail below, but the basic idea is that by planting or looking after forests, land-owners receive income in recognition of the environmental 'good' their forests perform. In this case, it's removing carbon dioxide from the atmosphere, but forests perform lots of other 'goods' too, including stopping soil erosion, keeping rivers clean and providing habitat for wildlife.

Our native forests are often overlooked in our climate change debates, but about 1.7 billion tonnes of carbon is stored in them, and scientists at the National Institute of Water and Atmospheric Research (NIWA) say that our forests could be 50 per cent better at storing carbon than we think.[5] The forests are not alone. Our soils, mangroves, saltmarshes and even tussock lands are all efficient carbon-storing machines, and we can help them to lock away even more by looking after them. Healthy plants that are not being eaten by possums, goats and other introduced species grow faster, storing more carbon along the way.

Forest destruction, on the other hand, turns our carbon sink into a carbon source. As the Parliamentary Commissioner for the Environment has said, a fire that destroyed all of our mature native forests would release 75 years' worth of emissions in one go.[6] This, of course, is the stuff of nightmares, and is hopefully something we will never see, but

---

4   https://www.mfe.govt.nz/sites/default/files/media/Climate%20Change/new-zealands-greenhouse-gas-inventory-1990-2018-vol-1.pdf
5   http://carbonnews.co.nz/story.asp?storyid=11443
6   https://www.pce.parliament.nz/media/1678/climate-change-and-agriculture-web.pdf

it highlights the fundamental point that forest protection is good and forest destruction is bad.

Reducing the amount of greenhouse gases in the atmosphere is called climate change mitigation. Learning to live with a changing climate is climate change adaptation. We have to do both, because as others in this book say, there are some aspects of climate change that are now unavoidable, no matter what we do.

Nature should be front and centre in our climate change adaptation plans too, because what's coming is too big to engineer our way out of with concrete and steel. Think about it this way: as the climate warms, our weather patterns will change. The heat will be more intense, the rains will be more intensive, the droughts longer. Healthy ecosystems will help lessen the impacts of all of them.

Imagine, for example, a drop of rain, hurtling from the sky in an intense storm. In many 'modified' landscapes it will slam into the ground, where it will join forces with its comrades to make a mighty flood, before eventually flowing into swollen rivers and out to sea, taking soil, nutrients, logs and whatever else it can carry with it, damaging everything in its path. We know this happens; we saw it at Tolaga Bay in 2018, when thousands of tonnes of 'slash', debris left behind after logging, was washed down to the coast in major flooding. If the raindrop falls on a forest, however, trees will slow its passage. Soil that's full of organic matter will soak up the raindrop like a sponge, holding it there for plants to use before it slowly trickles its way through the ground, into a stream or river and out to sea.

Trees also keep the temperature down, as anyone who has sat under a tree on a hot day knows. Cities, where dark buildings and roads absorb solar energy and release it as heat (a phenomenon known as the urban heat-island effect) are often several degrees warmer than the surrounding countryside. Replacing paving and other solid structures with nature (grass, shrubs, trees) cools the city, helps stop flooding, reduces air pollution (by trapping and dispersing particulate matter) and, of course, contributes to the vital work of pulling carbon dioxide out of the atmosphere.

Cities around the world are starting to catch on. Paris is creating an urban forest and making half its surfaces 'vegetated and permeable'. In the Netherlands, Arnhem is digging up paved roads and creating shady places around its shopping centres. Adelaide has found that watering city parks during heatwaves can cut temperatures by a whopping 12°C, and in the Philippines, plans have been unveiled for a 115 metre tall, 32-storey, all-wooden apartment building that incorporates 30,000 plants in the design.

At our coasts, dune systems, estuaries, mangroves and saltmarshes can form a buffer against rising oceans, but only if they are allowed to shift inland as the sea encroaches, as they have done over millions of years of changing sea levels. This time, though, they will run into us in many places. We like living near the sea — most of our biggest cities perch around harbours — and we like to think we can protect our homes and businesses with strong seawalls. But the energy involved in the oceans and weather is much bigger than us, and ultimately the seas will win.

Parliamentary Commissioner for the Environment Simon Upton addresses this in his recent report on estuaries: 'In the short term, it might make sense for communities to decide to make small upgrades to their flood protection schemes,' he says. 'But doing so may create a perception of safety and lead to ongoing investments and development, which then require further protection. This is the road of path dependency.'[7] In other words, we risk getting stuck on the path we're on. The best option, Upton says, is to start planning now so we have lots of time to manage sensible, orderly retreat from our coastlines. Examples might be making it mandatory for new buildings to be constructed in such a way that they can be moved relatively easily, and establishing a public fund to help pay

---

7   https://www.pce.parliament.nz/publications/managing-our-estuaries

for the cost of relocating large numbers of our citizens.

It is an issue the insurance industry is acutely aware of, too. With billions of dollars at stake, it's got a vested interest in seeing New Zealand adopt the most effective, least expensive options. It's pulling no punches, either. In 2016, Insurance Council chief executive Tim Grafton told a conference in Auckland that climate change would burst the Auckland housing bubble.[8] As sea levels rose, he said, insurance companies would stop offering long-term insurance on properties in low-lying but expensive suburbs such as Mission Bay and Kohimarama, and that would affect the ability of many potential buyers to get mortgages. Grafton advocated so-called soft-engineering options, like sand dunes and mangrove forests, as one of the best ways to protect the city.

What he didn't address, however, is who is going to pay for all this, and that *is* a problem. For years we've exploited the benefits of nature — those ecosystems services — without paying our way. If mangroves and sand dunes are needed to protect Auckland from rising seas, who should pay for them? Taxpayers? Ratepayers? The owners of properties sheltering behind them? Insurance or finance companies? A new public fund, as recommended by Upton?

Our brains are capable of coming up with lots of ways of figuring this out, and politics will determine whether we go with the user-pays or public-good way of doing things. What is critical, though, is that we realise these places have 'value' and make sure they are looked after.

The threat of rising seas is not solely an urban problem, either. Some of our 'best' farmland, on the Hauraki Plains, is on nearly 100,000 hectares of drained wetland. Until the early twentieth century, the region was unsuitable for farming. But then the government built stop-banks along the rivers and criss-crossed the plains with canals. Throw in a massive pumping system and you've got prime dairy country — a classic example of our belief that we can control nature.

---

8   http://carbonnews.co.nz/story.asp?storyid=11281

There are two problems, though. One is that much of that land is peat, which is releasing methane — a potent greenhouse gas — into the atmosphere. The other is that as sea levels rise, the pumps that currently keep the plains drained will no longer be up to the job. At the same time, rising seas in the Firth of Thames will mean the migratory birds and other species that use the mudflats and shell banks at Miranda — a site recognised with international Ramsar status for its global ecological importance — will be trying to move inland.

The Waikato Regional Council has come up with a plan to deal with this by working with nature. It wants to use 'blue carbon' (storing carbon dioxide in coastal ecosystems like mangroves) and paludiculture (wet-farming systems like those used for buffalo, in which peatlands are waterlogged once again to stop carbon from binding with oxygen and escaping into the atmosphere). It also wants the carbon stored in mangroves and saltmarshes officially recognised in the country's carbon accounts — a sentiment echoed in New Zealand's current Biodiversity Strategy.[9]

Ideas like this might sound a bit revolutionary — maybe even crazy, if you are the sort of person who's deeply wedded to the way things are — but revolutionary change over the coming decade or two is the one thing we can be completely sure of. Because of what we have done (i.e. burn fossil fuels and clear land) since the 1870s — and especially in the past 30 years — there is now no way for us to avoid major change. What we can control, however, is the speed and scale of that change, and whether we allow it to take a terrible toll, or we seize opportunity with both hands and build a better world.

---

9   https://www.doc.govt.nz/nature/biodiversity/aotearoa-new-zealand-biodiversity-strategy/#:~:text=Te%20Mana%20o%20te%20Taiao%20%2D%20Aotearoa%20New%20Zealand%20Biodiversity%20Strategy,Zealand%2C%20from%202020%20to%202050.

The new normal

Sometimes it's like a giant hand takes our world and shakes it. When it stops and we can look around, we find some of the old pieces are still there, but they're arranged in a different way and, somehow, we need to use them to build a new way

Our thinking about climate change has much of the magical about it.

of living. We're not always good at this; we cling to what we know and, for a while at least, try to carry on as if nothing's changed, even though a rational reading of what's in front of us says it has.

In her book about the year following the sudden and unexpected death of her husband, American novelist Joan Didion called it a kind of magical thinking. She knew her husband was dead — she'd called the ambulance, had seen the paramedics fail to revive him — yet she kept expecting him to return, to the point where she wouldn't get rid of his clothes in case he needed them.

Our thinking about climate change has much of the magical about it. It is unreal, we think; just a hoax. It can't possibly happen in our lifetime. If it was that bad, someone would be doing something about it (actually, that thought isn't unreasonable, but sadly it's not the case). Or we think it *is* real, but switching to electric cars, eating less meat or adopting some yet-to-be-invented technology will fix it. Otherwise, life will go on as normal.

History is full of points where the world is reshaped and the lives people expected to live are jolted onto another trajectory altogether. Our world is vastly different from that of our ancestors because of industrialisation, colonisation and globalisation. The world they lived in, which was real and solid to them, is just a story to us, something we learn about but never experience.

The Covid-19 pandemic has shown us just how fast 'normal' can be swept away. At the time of writing, New Zealand is in its second lockdown. My husband is seriously ill, but his children and grandchildren, who live just 160 km away, cannot come and see him because no one is allowed to leave Auckland. Six months ago, that was unthinkable. A trip to Auckland took a couple of hours and we could do it whenever we liked. We assumed that would always be the case, and yet suddenly it isn't.

Of course, we expect this to end. Aucklanders will be allowed off their mainland island when it's deemed they no longer pose a threat to the rest of us, and life will get back to normal. But it will be a *version* of normal, not the one we expected to live with our whole lives.

As I write, our country's borders seem likely to remain closed for the foreseeable future, starving the tourism and educational industries of their international customers and denying us our mid-winter holidays in the sun. A generation faces the prospect of being unable to fulfil the Kiwi rite-of-passage, the Big OE, and families with members stretched across the globe can no longer count on regular get-togethers courtesy of cheap, fast air travel. It is now harder to travel than it was in the nineteenth century. Who'd have thought?

Experts. Epidemiologists. Public health officials, that's who. Possibly someone in Treasury who had run the numbers on what a pandemic could do to the economy. They all knew that it could happen. The rest of us trundled along in our own little worlds, not thinking about what a pandemic could do to our lives until it was completely and utterly unavoidable.

Unfortunately, that has pretty much been the story of climate change, too. Scientists have known about it since the 1890s. It came to the attention of environmentalists and environmental policymakers in the 1980s. But it is only now, when the world is experiencing record heat, droughts, floods and, yes, a pandemic, that the general population is starting to think that this might just be a problem we should do something about.

If only we'd got to this position about 30 years ago, when there was still time for a don't-scare-the-horses transition to a low-carbon world. In 1990, New Zealand's annual emissions stood at 63.59 million tonnes.[10] When carbon stored in trees that year was taken into account, our net emissions came up out at 35.29 million tonnes. In 2018 (the most recent year for which figures are available), our gross emissions stood at 78.86 million tonnes and our net emissions were 55.49 million tonnes — a 57 per cent increase, despite the fact that we made an international commitment that by 2020 our emissions would be 5 per cent lower than they were in 1990.

Think about that for a moment. We New Zealanders, among the richest people in the world, have, in the 30 years since we started participating in international negotiations to lower greenhouse gas emissions, increased our greenhouse gas emissions by more than 50 per cent instead of lowering them.

The rise is down to three things: methane from the agricultural industry and carbon dioxide from the transport and forestry sectors. In other words, cows, cars and cutting down trees. Which makes those three things a good place to start if we want to reverse the trend.

It's all about attitude

But before we look at specifics we need to think about our attitude, because attitude is the single most important thing we have to change. As with the Covid-19 pandemic, change is not optional. We cannot continue life as usual. If we try to party on like there's no tomorrow, we will have to cope with changes to the physical environment that are probably beyond our imaginations right now.

Already we are locked into some level of climate change that we cannot now avoid. The only choice we have is whether we pull every

---

10   https://www.mfe.govt.nz/publications/climate-change/new-zealands-greenhouse-gas-inventory-1990-2018

> **If we try to party on like there's no tomorrow, we will have to cope with changes to the physical environment that are probably beyond our imaginations right now.**

lever we've got and limit the damage (recommended), or do nothing and live with the consequences. It is essential that we grasp this and factor it in to every piece of our decision-making, from whether it's a good idea to build a new hospital on low-lying land next to a river (it's not) to the kind of jobs we're preparing our kids for, the diseases we are going to have to deal with and what crops we should be planting now for future harvest.

There are good lessons to be learnt from the pandemic. New Zealand's 'go early and go hard' approach meant that as soon as those in charge (the government and health officials) knew what was happening, they took immediate, drastic steps to prevent the virus spreading. We were largely confined to our homes, and vast sections of the economy were shut down. The trade-off, we were told, was that decisive action now meant we'd be better off later, which we accepted.

Imagine if such decisive leadership on climate change had been shown in 1990 (or in any of the three decades since then). Imagine if it is shown now.

Sizing up the job

As Dr Jim Salinger says earlier in this book, New Zealand's temperatures have already warmed by at least 1°C in the hundred or so years we've been keeping records. He and other scientists have told us very, very clearly that we need to keep warming to no more than 1.5°C if we want a world that's relatively safe for us and most other species to live in.

So there's not a lot of room left to move. If the world carries on emitting carbon at the current rate, we will be at 1.5°C of warming by

2030 (it's worth remembering that back in 1990, average temperatures were only 0.5°C above pre-industrial levels, so most adult New Zealanders have already experienced substantial climate change in their lifetimes).

Currently, New Zealand's official target under the Paris Agreement on climate change is that by 2030, our emissions will be 11 per cent lower than they were in 1990. But this isn't going to do the job; if every country adopted a similarly unambitious target, we'd have warming of 3°C by the end of the century. What we're really looking at is a halving of emissions by 2030 to put us on the road to net-zero emissions by the middle of the century, and then some time in negative emissions to get things back on a more even keel.

What does that look like on the ground? Emissions reductions of around 6 or 7 per cent a year should get us there. For New Zealand, that means increased reductions of 3 million to 4 million tonnes every year. (And a note here on the difference between gross emissions and net emissions: gross emissions are *all* the greenhouse gas emissions New Zealand is responsible for. Net emissions are the gross emissions *minus* the carbon dioxide removed from the atmosphere that year and stored in New Zealand's forests and soils. The net emissions are what's important, because that's what's actually in the atmosphere, doing the damage.)

How big an ask is that? Big, but not impossible. And that's what is really important to remember: *we can do it*. It is physically possible. The hardest part is *deciding* we're going to do it, and then actually starting.

Think global, act local — together

It's hard to take the bus to work if there's no public transport system in your town. Or the last bus leaves before you finish work. Or you need to get groceries on the way home but you can't because the buses you need don't connect. In other words, we can't leave this job up to individual responsibility.

Sometimes you hear people saying that because climate activists got to a demonstration in petrol-driven cars that's a reason not to listen to them. It's a spurious argument, like a children saying that an inconsistency in one thing said by their parents means all other advice from them can be safely ignored.

We operate within a system that limits our ability as individuals to change the world. The things we can do are important, but saving our collective neck requires collective action on a mass scale. Instead of looking for reasons to blame each other and excuse ourselves, we need to work together to come up with brilliant ideas for new ways of doing things, and then have the guts to work together to make them happen.

We operate within a system that limits our ability as individuals to change the world.

It is tempting to play the blame game. Townies say farmers and their belching cows are the problem. Farmers say it's townies and their belching SUVs. Both are right (agriculture and transport are New Zealand's two biggest sources of emissions) and both are wrong, because New Zealand's emissions are a complicated, intricate web of cause and effect in which we are all — every single one of us — implicated.

Until recently, our emissions figures were based on production; in other words, we counted the emissions actually released within the country, regardless of where goods associated with those emissions were originally produced or finally consumed. So the emissions from operating our cars and trucks and factories and power stations and rubbish dumps and coal mines and commercial forests and, yes, our farms were included, but the emissions involved in the *production* of the goods we imported were not. They were attributed to China or Mexico or wherever else our televisions, cars and mountains of imported plastic goods were made.

But in 2020, Statistics New Zealand produced our first set of emissions accounts based on consumption.[11] It is illuminating reading. Yes, our production emissions are higher than our consumption emissions, because we export more than we import. But that doesn't mean our imports don't matter. In 2017, manufactured goods consumed by households were responsible for 22.4 million tonnes of emissions — more than twice the amount of emissions our entire manufacturing sector was responsible for producing the same year (10.7 million tonnes), and more than half the combined total (41.3 million tonnes) of the farming, fishing and forestry sectors.

The consumption accounts are a picture of just that — our growing consumption. After taking a hit from the Global Financial Crisis in 2007, our consumption emissions have been growing steadily each year and keep hitting new highs, despite a fall in what is called emissions intensity, a system of measuring emissions efficiency. That means that while the production of each individual widget we consume has led to the release of fewer greenhouse gases, any gains have been undermined by New Zealand's growing consumption and population. In other words, there are more of us, and we're consuming more.

Live within our means

And therein lies the lesson. We need less stuff. A *lot* less stuff. Possibly not a popular notion when the economy has already been hit for a six by the Covid-19 pandemic, but it's the truth. The idea of endless growth — of producing and consuming more and more and more — runs smack into the reality of planetary boundaries: the physical limits to the planet we live on, such as how much freshwater there is and the amount of greenhouses gases the atmosphere can absorb without the climate going haywire. Just as we can't go on forever spending more money than we

---

11   https://www.stats.govt.nz/information-releases/greenhouse-gas-emissions-consumption-based-year-ended-2017

earn, neither can we use more resources than the world has. Eventually there comes a reckoning, when the bank demands to be paid.

The only way to avoid this mortgagee-sale-by-the-physical-world is to live within our means. And that means cutting emissions to a level the atmosphere can tolerate.

It would be wonderful to say here that technology will do the job but, at this point at least, that would be grossly misleading. The only way to stop the world as we know it from being destroyed by dangerous levels of climate change is to stop emissions being released, and increase the amount of carbon dioxide being pulled out of the atmosphere and stored. And that means we do two things, urgently: stop burning fossil fuels, and look after nature. Because there are no technological silver bullets on the horizon to save us. We have to work with what we've got in front of us now.

I've already talked about looking after nature, and unless you're particularly hard-hearted, you probably agree with the concept, at least. What's not to like about greening our cities and keeping our wild places healthy and vibrant? Kicking the fossil-fuel habit is much more challenging, but that's where we especially need to pull together, by making it easy to do what needs to be done. And that doesn't mean supplying everyone with an electric vehicle, because although New Zealand's electricity is generated largely from renewable sources, the idea of every New Zealander who's old enough to hold a driver's licence owning and using a car every day is one drenched in the unsustainable consumption of planetary resources.

What we need to do is think about *efficiency*. If we are to keep warming below 1.5°C, there is a finite amount of greenhouse gases that we can emit into the atmosphere. This is called a carbon budget. As with all budgets, we need to think about the best way of spending — our emissions, in this case — so that what we have goes as far as possible.

It makes no sense, for example, to spend emissions on shipping carrots or oranges from the other side of the world when we grow excellent carrots and oranges in New Zealand. Coconuts and coffee don't grow here, however, so by limiting imports to these two items and encouraging local growers to supply the domestic instead of export markets, we have (by a very rough rule) halved our emissions.

When we take another look at it, we see there are still emissions involved in transporting our locally grown oranges and carrots around the country. But do all oranges have to come from Kerikeri, and all carrots from Ohakune? Carrots grown in your own backyard have the lowest transportation emissions of all (the energy it takes you to pull them from the garden and carry them to the kitchen). Just a bit above that are carrots grown by horticulturalists in your neighbourhood and sold in local shops or in farmers' markets.

Oranges are a little harder, because at the moment they do well only in our warmest regions, but there are still questions we can ask ourselves which could lead to a reduction in emissions. The first is whether we actually *need* oranges, or will the grapefruit and lemons that grow pretty well anywhere in New Zealand give us the vitamin C hit we need? The second is how the changing climate is going to affect horticultural production. With New Zealand experiencing its warmest winter on record in 2020, and seven of the ten warmest winters

What we're talking about here is rearranging our systems: food, transport, work, recreation ... everything.

occurring in the past 20 years, it's likely we are going to be able to grow oranges in places we haven't been able to grow them before.

What we're talking about here is rearranging our systems: food, transport, work, recreation . . . everything. By rearranging our food system with an emphasis on feeding our own communities first, we can cut emissions as well as create local employment, improve the health of our people through a combination of gardening and eating fresh food, and increase our resilience.

Communities that produce their own food have a certain

independence; they don't need the rest of the world to help them do that most basic of human activities: eat. How much stronger are communities that, in the event of, say, having their road and rail links to the rest of the country taken out by storms (an event that is going to happen more often as the impacts of climate change kick in), can at least feed themselves? And it's not just the consumers of food who benefit; during the first Covid-19 lockdown, exporters were rushing to find ways of selling their produce locally. (Producers are catching on to this idea; accounting firm KPMG's 2020 Agribusiness Report says that farming leaders want to shift the focus away from exporting towards feeding New Zealanders first.)

In the transport sector, we can reduce emissions from the operation of vehicles by switching to electric cars, and we can cut them even more by leapfrogging right over private vehicle ownership altogether if we put our minds to it. Is having a million Aucklanders spending several thousand dollars each a year to own cars so they can then sit trapped in them on congested roads for several hours a day just to get to work really the best, most efficient way to do things? Or is it better to arrange our cities so that people can live and work in the same community most of the time (by working from home, working from local shared 'hubs' or increasing the availability of affordable, comfortable housing in the central city), or have access to fast, effective and cheap public transport systems — including car-sharing or rental schemes — for those times when they want or need to go somewhere?

Do we really need Air New Zealand, a company which is majority-owned by the government, competing with and beating other forms of transport as the best way to move people in and out of our major cities every day, when aviation was responsible for 3.5 per cent of global emissions (according to 2103 figures), and emissions from aviation have more than doubled over the past 20 years?[12] Or could

------

12   https://www.sciencedirect.com/science/article/pii/S1352231020305689?via%3Dihub

we do things better — and more emissions-efficiently — by beefing up our inter-city rail service (also state-owned), and making it efficient and comfortable enough to spit people out in Auckland or Wellington in a fit state to work?

And when we find these efficiencies, have we got the strength of character to let them actually deliver the emissions reductions we so desperately need to find, or will we use them as an excuse to do more of what we've been doing? Hopefully our performance to date isn't a sign of things to come. As well as the wasted improvements in emissions efficiency in the consumer sector I've already mentioned, we've also undermined the benefits of greater public transport use in Wellington and Auckland on weekdays by driving further and in bigger cars at the weekend. And in the agricultural sector, instead of using a 20 per cent improvement in emissions efficiency to cut emissions, we've used it to increase the stocking rates of our farms.

Again, it comes down to our collective will, and that's something that can be influenced, by everything from public relations to regulation and markets. Pulling out of this mess is going to take all three approaches.

Advertising, marketing and public relations are all part of the art of persuasion — the very thing that for decades has convinced us to buy more and more. Their power to do good as well as bad should not be underestimated; look what happened to throw-away plastics once David Attenborough showed

What we need around climate change is clear, consistent and accurate information about what's happening and why, and the plan to deal with it.

us an albatross trying to feed bits of plastic to a chick. Single-use plastic bags are banned and plastic straws are about to go the same way. On their own, these are not measures that will save the world, but they demonstrate what can happen when critical numbers of people get an idea lodged in their heads.

What we need around climate change is clear, consistent and accurate information about what's happening and why, and the plan to deal with it. Not propaganda: *information*. We also need our leaders — political, business, community — to get informed and to speak honestly and accurately about climate change. We have a tendency to believe — not unreasonably — that if a threat is big and real, our leaders will be taking it seriously, as happened with Covid-19. However, the flip side of this coin is that if our leaders are *not* seen to be taking it seriously, it must mean that we don't have to either.

Persuasion alone will not be enough; we also need rules and regulations. Despite often being branded as 'red tape' or 'the nanny state', rules give everyone the clarity they need to get on and make decisions. If, for example, Simon Upton's suggestion is picked up and regulations are made requiring all new coastal buildings to be on stilts and relatively easy to move, we will have clear information about the future of what we're building, and we won't be kicking the cost of dealing with it down the road.

There are lots of things we can usefully make rules about, such as setting limits on vehicle emissions (New Zealand is one of the few countries in the world that don't have these already), requiring fossil fuels to be blended with increasing amounts of low-emissions fuels, transparency about the amount of money organisations such as our private superannuation funds have invested in high-carbon industries like oil and gas, and increasing our energy efficiency (that word again) by making things like effective insulation and double glazing mandatory in all buildings.

And then there's the market.

A real price on carbon

There's a concept in economics called 'the tragedy of the commons'.[13] The 'commons' are anything that is jointly, or commonly, owned, but I always get a mental picture of a Wombles-of-Wimbledon-type expanse of land that everyone is free to use. Because no one particular person owns the common, there is no incentive to use it wisely. In fact, if you don't take advantage of the flush of grass on the common while you can and get your animals fat by eating it all, someone else will. Then there will be none left for you or your animals — a tragedy for you, your animals and every living organism that relies on that grassland.

Climate change is a classic tragedy of the commons. No one owns the climate; it's just there, providing critical support for every single piece of life on Earth and totally ignoring any boundaries imposed by humans. Because of that, there's no incentive for any one person — or even country — to take care of it.

Scientists have known for more than a hundred years that burning fossil fuels on a massive scale will damage the climate, but business and political leaders have failed to do anything about it because they fear it will cost them money or votes. So we carry on discharging greenhouse gases in ever-increasing amounts, because if we don't, someone else will and will get the profits that could have been ours.

It's a daft system because, of course, all of us bear the costs in the end. But if we don't want to live in a state where we are simply told what to do, we have to find another way of encouraging the behaviour we want and discouraging the behaviour we don't want. That's where carbon pricing comes in. Imposing a cost on environmental damage — what economists call 'internalising the externalities' — should, if done properly, even things up by creating a disincentive to pollute.

---

13   Hardin, G., The Tragedy of the Commons, in *Science*; 1968, vol. 162, no 3859, pp. 1243–1248.

So how does it work? There are two types of carbon pricing: a carbon tax, which is pretty self-explanatory (you're taxed for whatever emissions you're responsible for, creating a very direct disincentive to pollute); and a market-based system known as a cap-and-trade scheme. New Zealand yo-yoed around with both until finally, in 2008, settling on the latter and passing legislation creating the New Zealand Emissions Trading Scheme.

Under a cap-and-trade scheme, a certain volume of emissions is allowed, based on how much more the atmosphere can take without causing dangerous levels of climate change. This is the cap part of the equation. Carbon credits or allowances are issued to represent each tonne of emissions up to the cap. It's up to each country to decide how they will use these emissions; at the moment in New Zealand we spend them mainly on subsidising agricultural and industrial polluters. Credits can also be issued for each tonne of 'removals' — carbon dioxide which has been taken from the atmosphere. Under our scheme, this is mainly carbon stored in trees as a result of photosynthesis, but it could be adapted to recognise other removals, such as carbon stored in soils or in silicate rocks or biochar if we figure out a way of using those systems at a significant scale.

This is where the 'trade' part of the equation comes in. Emitters with carbon liabilities can buy credits from forest owners and others who have been given allowances, and surrender them to the government.

At the heart of it, it's quite a simple idea. People who do things that remove carbon dioxide from the atmosphere or prevent greenhouse gas emissions that would have otherwise happened get credits. They can sell them to people responsible for emissions, who then surrender them to the government as a way of 'paying' for their emissions. The credits are then cancelled by the government so they can't be used again.

So, for example, the owner of an industrial plant which burns coal for heat and emits 1 million tonnes of greenhouse gases in a year buys credits from someone whose forest has sequestered 1 million tonnes of carbon in a year. From a climate change point of view, the forest cancels out the industrial plant, a situation known as net-zero emissions.

In theory, the price for the units comes down to good old supply-and-demand. As the annual cap decreases, so will the supply of allowances, pushing up the price of carbon credits and creating an incentive for the forest owner to plant more trees. Eventually, the price of credits will get so high that it is too expensive for the plant owner to keep burning coal, and other, low-carbon renewable forms of energy such as biofuels or renewable electricity become viable.

New Zealand might have had an emissions trading scheme since 2008, but it has abjectly failed to find this pricing sweet-spot. Artificial controls on prices and demand introduced by the National government in 2009, along with a total lack of control on supply (emitters were allowed to 'import' cheap, environmentally worthless carbon credits), saw the value of units being used in the New Zealand ETS fall to a pathetic 17 cents. Add that to the subsidies in place at the time, and our industrial-plant owner was looking at a carbon bill of just $8500 a year — hardly enough to attract the attention of the top brass, let alone make the share market quake.

By mid-2020, some subsidies had been removed, others were on the way out, the price of carbon was around $34 a tonne and our friend the plant owner had a much more attention-grabbing carbon bill of $3.4 million per annum. Suddenly, finding a low-carbon source of heat made more financial sense.

Imagine, then, what complete removal of subsidies and a carbon price of, say, $200 a tonne looks like: a $200 million annual carbon cost. That's not a figure plucked out of the air; heavyweight organisations such as the World Bank say carbon prices are going to need to reach that kind of magnitude if they are to have a real effect on emissions.

Discussions about carbon markets tend to have one of two effects on people: they are either intensely bored, or they become angry because the ETS has failed to drive any emissions cuts in the more than a decade it has been in place, and they consider it to be more of the same type of thinking that brought the climate crisis upon us in the first place. But throwing it out now would be a mistake — mainly because it's in place, and there is broad political acceptance of it. Anyone who lived through the years and years of political shenanigans it took to get this far will throw their hands up in horror at the thought of having to start over.

And, quite simply, we haven't got time. By the time this book is published, we will have just nine years in which to halve our emissions and do our bit towards keeping warming below 1.5°C. We have left taking action for so long that now we have to do everything we can, as quickly as we can.

Price carbon properly by making it reflect the true cost of the damage greenhouse gas emissions are causing. Go for the low-hanging fruit first — that's transport and forestry (especially native) — but don't forget that the harder staff like agriculture still has to be done. Understand that our way of living is changing whether we like it or not, and embrace the chance to do things in better ways. Invest in exploring new technologies, but don't count on a silver bullet coming to save us. And, above all, learn to live with nature, instead of fighting against it.

Author
biographies

Rob Bell

Rob has 40 years' experience in coastal engineering, risk from natural hazards, the impacts of climate change on coastal communities and infrastructure, and how we can adapt. Until recently, Rob was the programme leader for NIWA's (National Institute of Water and Atmospheric Research) climate-change adaptation mahi. Rob was the lead author of the 2017 coastal guidance for local government published by Ministry for the Environment for planning adaptation to climate change. He is currently a contributing author for the IPCC Working Group II sixth assessment report on climate change impacts for Australasia. Rob received Lifetime Achievement Awards in 2019–20 from NIWA, Science NZ and KuDos (Waikato) and was part of the Victoria University New Zealand SeaRise team awarded the Prime Minister's Science Prize in 2019.

Jason Boberg

Jason Boberg is a proudly disabled advocate and social entrepreneur. The founder of disability climate network, SustainedAbility, he has worked in the United Nations Convention on the Rights of Persons with Disabilities and the UNFCCC to highlight the impacts of climate change on disabled people, and to promote the establishment of a formal disability constituency. As the co-founder of social impact agency, Activate Agency, he brings a critical disability rights lens to his work, focusing on ethical representation of disabled people in media, governance, and social change; and provides training and policy advice to NGOs, DPOs and government in Aotearoa and abroad. Jason is a Climate Reality Leader and is a member of the Auckland Council Disability Advisory panel and is on the Advisory Panels' Climate Change Working Group.

Helen Clark

Rt Hon. Helen Clark is a former New Zealand Prime Minister and a former UNDP Administrator. She currently chairs the boards of the Extractive Industries Transparency Initiative and the Partnership for Maternal, Newborn, and Child Health. Helen also serves on other public good advisory boards. Helen is a frequent contributor to conferences, and events, on issues related to sustainable development and the Sustainable Development Goals (SDGs) and Women's Leadership.

Helen is currently co-chair of The Independent Panel for Pandemic Preparedness and Response mandated by the World Health Assembly to review the internationally coordinated response to the Covid-19 pandemic.

Adelia Hallett

Adelia Hallett is an environmental journalist who has been covering climate change for more than 30 years. With her late journalist husband, Sandy Macdonald, she developed the Carbon News daily news service. She has also worked for major news services including *The New Zealand Herald* and Radio New Zealand's Mediawatch programme, and for a range of government departments, environmental and social organisations and private businesses. She lives in Northland, where the view from her desk includes a marine reserve and the Marsden Point oil refinery — two reminders of both the problems and the opportunities of tackling climate change.

Sophie Handford

Sophie Handford is a 20-year-old councillor and activist from Kāpiti, New Zealand. Since the age of 12, Sophie has held the environment close and has always felt a strong connection to and need to protect the planet we share. Sophie was service captain, student rep and head girl at various points throughout her secondary schooling at Kāpiti College. After graduating, she founded School Strike 4 Climate in Aotearoa and went on to coordinate the movement which mobilised 170,000 people across the country.

Andrew Jeffs

The New Zealand coast has fascinated Andrew Jeffs since he was a child who spent his summers exploring rockpools and catching spotties and sprats off wharves. Today he is a marine scientist at the University of Auckland who studies on a wide variety of coastal issues including the restoration of mussel beds, the farming of crayfish, seaweed and mussels, and the feeding of seabirds.

Rhys Jones

Dr Rhy Jones, Ngāti Kahungunu, is a public health medicine specialist and senior lecturer in Māori Health at the University of Auckland. His research interests include ethnic inequalities in health, Indigenous health in health professional education, and environmental influences on health. In 2005–06, he was a Harkness Fellow in Health Care Policy at Harvard Medical School, examining interventions to reduce racial and ethnic disparities in health care. Rhys was the founding co-convenor of Ora Taiao: The New Zealand Climate and Health Council, a health professional organisation focusing on the health challenges of climate change and the health opportunities of climate action.

Haylee Koroi

Haylee Koroi is from the inland Utakura valley and the small coastal settlement of Pukepoto in the Far North. Her movements are guided by a philosophy that realigning with whakapapa is key to the wellbeing of all of our human and more-than-human relations. This has led her to work in spaces of Indigenous education and training, Māori public health, climate justice and creating alongside other Indigenous women. Regardless of spaces occupied, she believes that if we are centring whanaungatanga in our practice then we are honouring the legacies passed down to us by our ancestors, and inherited from us by our grandchildren.

Matt McGlone

Matt McGlone is an emeritus researcher based at Manaaki Whenua-Landcare Research, Lincoln. His main research interest is the vegetation and climate history and plant biogeography of Aotearoa New Zealand. In 1989, he was the lead author of the book *Unsettled Outlook*, an early examination of the approaching climate crisis. In the course of his career he has provided scientific advice on biodiversity and climate change policy to a number of government and international agencies. Most recently, he has assisted the Department of Conservation in the design and implementation of a comprehensive biodiversity monitoring scheme.

Jamie Morton

Jamie Morton has been a journalist for nearly two decades — much of which has been devoted to covering science and environment issues for *The New Zealand Herald*, the country's largest circulating daily newspaper. Morton has travelled to international disaster zones, the landmark 2015 UN COP21 climate change conference in Paris and twice to Antarctica, where climate science matters most. Morton has won several major media awards for his reporting.

Rod Oram

Business journalist Rod Oram contributes weekly to *Newsroom*, Nine to Noon and Newstalk ZB. He is a public speaker on deep sustainability, business, economics and innovation. Rod is a member of the Edmund Hillary Fellowship which brings together innovators and investors from here and abroad who seek to contribute to global change from Aotearoa. In Citigroup's annual global journalism awards Rod was the winner in 2019 in the General Business category in the Australia and New Zealand region for his columns in *Newsroom* on Fonterra, and he was the New Zealand Journalist of the year. In the 2020 and 2018 New Zealand Shareholders' Association Business Journalism Awards, Rod won the Business Commentary category for his *Newsroom* columns.

Jim Salinger

Dr Jim Salinger is a climate scientist of international repute, the first to uncover climate warming in the New Zealand region in the 1970s. He contributed to the Nobel Peace Prize (2007) together with Intergovernmental Panel on Climate Change (IPCC) colleagues for ground-breaking work on climate change. He held several visiting professor positions including Stanford and University of Florence (Italy). Jim is an inexhaustible publisher in academic journals and prolific communicator of climate change to the public. He was recently awarded the Jubilee Medal, the premier award from the New Zealand Institute of Agricultural and Horticultural Science, for lifetime achievements in climate and agricultural science.

Kera Sherwood O'Regan

Kera Sherwood-O'Regan, Kāi Tahu, is the impact director at social impact creative agency, Activate Agency. With a background in political science and public health, Kera has over 15 years' experience in climate campaigning. She is a Climate Reality Leader, founding member of SustainedAbility, and former board member of OraTaiao: The New Zealand Climate and Health Council. Kera works annually at the United Nations Climate Negotiations supporting the International Indigenous Peoples' Forum on Climate Change. As a proudly disabled and Indigenous woman, Kera's work centres on structurally oppressed communities in social change, and crosses the intersections of indigenous & disability rights, health and climate change.

Simon Thrush

Professor Simon Thrush is director of the Institute of Marine Science and director of the George Mason Centre for the Natural Environment at the University of Auckland with research interests in marine ecology, marine ecosystem services, resilience and tipping points in marine ecosystems and human impacts on the environment.

Professor Thrush obtained a BSc (Hons) from the University of Otago and a PhD from the University of East Anglia, England. He has over 30 years' experience in the development and implementation of strategic ecological research to influence resource management and improve societal valuation of marine ecosystems. He has worked in New Zealand, Europe, USA and Antarctica, has contributed to over 200 publications in peer reviewed scientific literature and collaborates with colleagues around the world.